Visual and Non-Visual Effects of Light

Occupational Safety, Health, and Ergonomics: Theory and Practice

Series Editor: Danuta Koradecka

(Central Institute for Labour Protection – National Research Institute)

This series will contain monographs, references, and professional books on a compendium of knowledge in the interdisciplinary area of environmental engineering, which covers ergonomics and safety and the protection of human health in the working environment. Its aim consists in an interdisciplinary, comprehensive and modern approach to hazards, not only those already present in the working environment, but also those related to the expected changes in new technologies and work organizations. The series aims to acquaint both researchers and practitioners with the latest research in occupational safety and ergonomics. The public, who want to improve their own or their family's safety, and the protection of heath will find it helpful, too. Thus, individual books in this series present both a scientific approach to problems and suggest practical solutions; they are offered in response to the actual needs of companies, enterprises, and institutions.

For more information about this series, please visit: https://www.crcpress.com/Occupational-Safety-Health-and-Ergonomics-Theory-and-Practice/book-series/CRCOSHETP

Visual and Non-Visual Effects of Light

Working Environment and Well-Being

Agnieszka Wolska, Dariusz Sawicki,
and Małgorzata Tafil-Klawe

CRC Press
Taylor & Francis Group
Boca Raton London New York

CRC Press is an imprint of the
Taylor & Francis Group, an **informa** business

First edition published 2021
by CRC Press
6000 Broken Sound Parkway NW, Suite 300, Boca Raton, FL 33487-2742

and by CRC Press
2 Park Square, Milton Park, Abingdon, Oxon, OX14 4RN

© 2021 Taylor & Francis Group, LLC
CRC Press is an imprint of Taylor & Francis Group, LLC

Library of Congress Cataloging-in-Publication Data

Names: Wolska, Agnieszka, 1963- author. | Sawicki, Dariusz, 1957- author. | Tafil-Klawe, Malgorzata, author.
Title: Visual and non-visual effects of light : working environment and well-being / Malgorzata Tafil-Klawe, Agnieszka Wolska, Dariusz Sawicki.
Description: First edition. | Boca Raton, FL : CRC Press, 2020. | Series: Occupational safety, health, and ergonomics | Includes bibliographical references and index.
Identifiers: LCCN 2020023962 (print) | LCCN 2020023963 (ebook) | ISBN 9780367444198 (hbk) | ISBN 9780367529529 (pbk) | ISBN 9781003027249 (ebk)
Subjects: LCSH: Lighting--Health aspects. | Light--Physiological effect.
Classification: LCC RC963.5.L54 W65 2020 (print) | LCC RC963.5.L54 (ebook) | DDC 615.8/31--dc23
LC record available at https://lccn.loc.gov/2020023962
LC ebook record available at https://lccn.loc.gov/2020023963

ISBN: 9780367444198 (hbk)
ISBN: 9781003027249 (ebk)

Typeset in Times
by Deanta Global Publishing Services, Chennai, India

Contents

Preface

Sunrise, sunset
Sunrise, sunset
Swiftly fly the years
One season following another
Laden with happiness and tears

Fiddler on the Roof

Jerry Bock, music, and Sheldon Harnick, lyrics

For ages, sunlight has participated in the development of all life forms on Earth. The micro-world and the daily cycle of plants and animals all succumbed to the light–dark rhythm. Over thousands of years human beings lived in accordance with this pattern. It set a natural order which is the strongest of the circadian regulators. The discovery and development of artificial light sources eliminated the workings of this physiological clock. Today the external world is full of light pollution. We are now looking for solutions which would integrate care for the natural environment, energy-saving, and lighting efficiency. At the same time, we should pay attention to the biological properties of our visual system, because the human eye does not always function at its best in conditions which are thought to be optimal from the point of view of technology.

The circumstances of modern life generated shift work, which has been legitimized as a natural situation. We know the medical consequences for shift workers (Shift Work Disorder (SWD)), but this does not mean we either can or want to give it up; 24-hour operations are a practical necessity for the modern industrial economy. At the societal level, SWD is associated with dramatically increased accident risk and thus with financial and emotional costs borne both by workers and all of society, by employers and the general public. This is why it is so important to keep looking for solutions which would make it possible to build lighting sources that are as friendly as possible to our eyes and to the higher levels of our central nervous system, while minimizing the consequences of disrupting biological rhythms.

The authors look at lighting holistically, in recognition of its permanent and stable role in our lives. Biology, physics, lighting engineering, and occupational safety and health meet in this monograph in an interdisciplinary work. We hope that such an approach will increase the number of readers who will find it of interest.

Acknowledgments

Wishing a special thank you to Marzenna Rączkowska for proofreading this book.

Acknowledgements

Authors

Agnieszka Wolska, Ph.D., D.Sc. (Eng), was born in 1963. She graduated from the Faculty of Electrical Engineering (specialization: Lighting Engineering) at the Warsaw University of Technology (Poland) in 1988. In 1997 she earned her Ph.D., and in 2014 her D.Sc. (habilitation) degree from the Technical University of Białystok, Poland in the field of lighting engineering.

Since 1989, she has been working at the Central Institute for Labour Protection – National Research Institute (CIOP–PIB) in Warsaw, Poland, where she has held the position of Associate Professor since 2014. She has been head of the Optical Radiation Laboratory at CIOP–PIB since 1999.

Her current work involves activities in the area of lighting engineering, especially as related to the influence of lighting on well-being and health (experimental studies concerning evaluation of visual fatigue, visual comfort, glare assessment and measurement, psychomotor performance, and alertness related to lighting of different parameters) and visual ergonomics, as well as hazards arising from optical radiation.

Her scientific achievements include more than 90 scientific publications, such as monographs, chapters in monographs, and scientific articles.

She is a member and the Polish representative of Division 1 "Colour and vision" of the International Commission on Illumination (CIE) (since 1994), a member of the Polish Ergonomics Society (since 2005), a member of the Presidium of the Polish Committee on Illumination (1994–2002, and from 2014 until now), and a member of the European network of specialists in the field of occupational safety and protection EUROSHNET (since 2001).

Dariusz Sawicki, Ph.D., D.Sc. (Eng), was born in 1957. He earned his M.S. in Electrical Engineering in 1981, his Ph.D. in 1986, and his D.Sc. (habilitation) degree in 2008, all from the Warsaw University of Technology, Poland.

From 1986 to 2011, he was Research Assistant and Assistant Professor at the Warsaw University of Technology, and since 2011 he has been Associate Professor there. Since 2017, he has been Head of the Measurement and Information Systems Division at the Institute of the Theory of Electrical Engineering, Measurement and Information Systems of the Warsaw University of Technology. His main interests include visual perception, measurement, and simulation in lighting technology, HCI, and computer graphics. For almost ten years he has been involved in glare measurement in indoor and outdoor workplaces. Currently, he conducts research on fatigue detection using signals from various sensors and EEG analysis. His scientific hobby is examining perceptual problems concerning geometric and color illusion in art, computer graphics, and photography.

He is the author of more than 100 scientific publications: journal articles, chapters in monographs, and conference papers. Dr Sawicki has served as the representative of Poland in the 8th Division of CIE (International Commission on Illumination, Division 8: Image Technology) in the years 2014–2017 and 2018–2022. He has been a member of IEEE since 1992 (Senior Member since 2012), a member of ACM since 2001 (Senior Member since 2011), and a member of the Polish Information Processing Society since 1987.

Małgorzata Tafil-Klawe, Ph.D., Med.Sc., Professor of Medicine, is a graduate of the Faculty of Medicine (specialization: human physiology) at the Medical University in Warsaw (Poland), she is head of the Human Physiology Department of Ludwik Rydygier Collegium Medicum in Bydgoszcz, Nicolaus Copernicus University in Toruń (Poland). In 1982 she defended her dissertation for the Ph.D., Med.Sc. degree, and in 1991 her D.Sc., Med.Sc dissertation at the Medical University of Warsaw. In 2006 she was awarded the scientific title of Professor.

Currently she carries out research in the areas of human physiology, i.e. chronobiology with elements of chronomedicine, sleep physiology, cardiovascular regulation, respiratory medicine as well as in clinical physiology, e.g. in regulatory mechanisms in sleep apnea syndrome and the function of the autonomic nervous system in CFS/ME (Chronic Fatigue Syndrome). She was also involved in studies related to shift work disorder and the influence of different colors of light on acute alertness.

She is a member of the Polish Physiological Society and President of the Polish Society of Chronic Fatigue Syndrome (CFS) Research.

Series Editor

Professor Danuta Koradecka, Ph.D., D.Med.Sc. and Director of the Central Institute for Labour Protection – National Research Institute (CIOP-PIB), is a specialist in occupational health. Her research interests include the human health effects of hand-transmitted vibration; ergonomics research on the human body's response to the combined effects of vibration, noise, low temperature and static load; assessment of static and dynamic physical load; development of hygienic standards as well as development and implementation of ergonomic solutions to improve working conditions in accordance with International Labour Organisation (ILO) convention and European Union (EU) directives. She is the author of more than 200 scientific publications and several books on occupational safety and health.

***

The "Occupational Safety, Health, and Ergonomics: Theory and Practice" series of monographs is focused on the challenges of the 21st century in this area of knowledge. These challenges address diverse risks in the working environment of chemical (including carcinogens, mutagens, endocrine agents), biological (bacteria, viruses), physical (noise, electromagnetic radiation) and psychophysical (stress) nature. Humans have been in contact with all these risks for thousands of years. Initially, their intensity was lower, but over time it has gradually increased, and now too often exceeds the limits of man's ability to adapt. Moreover, risks to human safety and health, so far assigned to the working environment, are now also increasingly emerging in the living environment. With the globalization of production and merging of labor markets, the practical use of the knowledge on occupational safety, health, and ergonomics should be comparable between countries. The presented series will contribute to this process.

The Central Institute for Labour Protection – National Research Institute, conducting research in the discipline of environmental engineering, in the area of working environment and implementing its results, has summarized the achievements – including its own – in this field from 2011 to 2019. Such work would not be possible without cooperation with scientists from other Polish and foreign institutions as authors or reviewers of this series. I would like to express my gratitude to all of them for their work.

It would not be feasible to publish this series without the professionalism of the specialists from the Publishing Division, the Centre for Scientific Information and Documentation, and the International Cooperation Division of our Institute. The challenge was also the editorial compilation of the series and ensuring the efficiency of this publishing process, for which I would like to thank the entire editorial team of CRC Press – Taylor & Francis Group.

***

This monograph, published in 2020, has been based on the results of a research task carried out within the scope of the second to fourth stage of the Polish National Programme "Improvement of safety and working conditions" partly supported – within the scope of research and development – by the Ministry of Science and Higher Education/National Centre for Research and Development, and within the scope of state services – by the Ministry of Family, Labour and Social Policy. The Central Institute for Labour Protection – National Research Institute is the Programme's main coordinator and contractor.

1 Introduction

Light Light. The visible reminder of Invisible Light.

T.S. Eliot

1.1 COMMON DEFINITIONS OF LIGHT

Light is an important regulator of physiology and behavior in all living entities. Its significance is enhanced by the fact that life on Earth is subject to alternating cycles of day and night (light and darkness) imposed by the rotation of our planet. For humans, the sense of vision plays a central role when interacting with the environment. Consequently, most definitions of light are related to one's visual response to this phenomenon. For example, according to the Merriam-Webster Dictionary [Webster 2019], "light" (as a noun) has three definitions:

- something that makes vision possible,
- the sensation aroused by stimulation of the visual receptors,
- electromagnetic radiation of any wavelength that travels in a vacuum with a speed of 299,792,458 m (about 186,000 miles) per second, specifically; such radiation that is visible to the human eye.

These three definitions do not relate to the non-visual response to light. Why?

Although for a long time it has been widely known that sunlight is one of the most important regulators of human physiological functions related to circadian rhythm, in-depth research into the biological mechanism of the non-visual effect of light began in the late 20th and early 21st centuries, with the discovery of a new photoreceptor on the retina called the intrinsically photosensitive retinal ganglion cell (ipRGC). A rapid development of research related to the human non-visual response to light was launched. It has been proved that light has the ability to change circadian rhythms, i.e. to change the time periods in circadian cycles, which may result in shifting the phases of physiological cycles. Light also affects a number of physiological reactions, such as the regulation of hormone secretion (e.g. it can inhibit the pineal hormone responsible for melatonin secretion at night), affects the level of body temperature, induces the pupillary reflex, raises subjective alertness, and, under certain conditions, changes the bioelectrical brain activity indicating sleepiness or alertness. That is why the non-visual effects of light are so important for human functioning.

1.2 VISUAL AND NON-VISUAL RESPONSE TO LIGHT

Retinal photoreceptors make it possible to gather information passed from the eye to the visual parts of the brain, which analyze and modify this information to form a representation of objects, working in a system of conventional image-forming vision (visual responses to light). Light enters the eye through the cornea, and then through the pupil, whose diameter is controlled by the muscles of the iris. Behind the iris lies the lens, which in conjunction with the cornea focuses the incoming light on the back of the eye, i.e. on the retina, which contains light-sensitive neuron photoreceptors: rods and cones. Photoreceptors are responsible for the process of phototransduction: the conversion of the energy of the sensory stimulus, the photons of light, into an electrical signal – action potential, transmitted and analyzed by the cells of the nervous system. Five general classes of neuron make up the retina. One class, retinal ganglion cells, uses its axons to form the optic nerve, which carries visual information from the eye to the visual brain centers.

However, the form of photodetection involved in the synchronization of biological processes with the dark/light cycle, i.e. non-image forming vision (non-visual responses to light), in which the basic information is light or darkness over time, seems to be far more ancient than image-forming vision, and has been widely discussed in recent years.

Although it was originally believed that all light signals for image- and non-image-forming vision began with rods and cones, the first suggestion that these classical photoreceptors do not account for the spectral sensitivity of the pupillary light reflex can be found as early as 1923. In 1980 it was reported that light regulated the levels of the neurotransmitter dopamine in rats' retinas with degenerated rods and cones. Over the subsequent years, several studies showed a shift of circadian rhythms according to the external light/dark cycle in rodless/coneless mice. At the end of the 20th century similar observations were made in humans: light was effective in entraining the circadian clock in blind people without impinging on their conscious perception. The next step was a discovery of a new photopigment, melanopsin, localized to a small subset of retinal ganglion cells which project to the hypothalamus, and in particular to the suprachiasmatic nucleus (SCN), the master circadian pacemaker and biological clock, as well as to other brain regions serving non-image-forming vision. Subsequent important studies described intrinsically photosensitive retinal ganglion cells (ipRGCs), whose cells and the melanopsin-expressing retinal ganglion cells were shown to be one and the same. Both of these differ from other retinal output neurons; they show autonomous phototransduction, mediated by the photopigment melanopsin. They are diverse and are now thought to comprise five types, with different physiological functions. Generally, these cells are typically associated with non-image-forming functions: circadian photoentrainment and the pupillary light reflex, but some subtypes also influence the visual function, suggesting interaction between image- and non-image-forming vision. Recent studies provide evidence that ipRGCs can contribute to our awareness of external light, pointing to an important functional overlap of the ipRGCs and conventional rod/cone systems.

Thus, it is not a surprise that the complex morphological and functional retinal organization makes it possible to consider the dual nature of human responses to light and lighting: visual and non-visual. These responses are especially important in the context of human exposure to artificial light at night and during the day or in shift work conditions.

1.3 LIGHT AND CIRCADIAN RHYTHM

Day–night cycles regulated by daylight control the human internal biological clock known as the *circadian rhythm*. The name comes from the Latin phrase *circa diem*, which is translated as "around a day". It lasts for about 24 hours and is a continuous rhythm based on our body's reaction to the presence or absence of daylight. The common presence of artificial light of different spectra, both during the day and at night, has a significant influence on our physiology, including circadian rhythm disruption.

The circadian rhythm of human physiological functions is controlled by two clusters of neurons called the suprachiasmatic nuclei (SCN), which are located at the base of the hypothalamus in close proximity to the intersection of the optic nerves (hence the name). Information on the alternation of day and night reaches the SCN via a visual tract through photosensitive retinal ganglion cells. In response to light, melanopsin is activated and information is transmitted, thanks to which SCN cells begin "measuring" the next day. In the suprachiasmatic nucleus, the path to sympathetic centers in the thoracic spinal cord begins. From here, further fibers exit into the pineal gland, which secretes melatonin. The lack of light is a signal for the pineal gland to produce this hormone, and thus prepares the human body for the sleep phase. In contrast, the presence of bright white light or monochromatic light with specific wavelengths in the range between 420 and 550 nm inhibits the secretion of this hormone and puts the body in a state of readiness (wakefulness). It has been proven that light with a length perceived as blue (between 450 and 490 nm) is most responsible for the direct non-visual effect.

1.4 LED LIGHTING AND POTENTIAL HEALTH HAZARD

In the last ten years light sources and technology have experienced a revolution. The new generation of light sources – light-emitting diodes (LEDs) – have become widely used in industrial and commercial environments, but also in non-industrial applications: in TVs, computers, smartphones, and tablets. The most common white LEDs used for illumination are phosphor-based ones. Usually, blue light LEDs (450 nm) are covered by phosphor material which converts monochromatic blue light into broad-band white light. But regardless of the phosphor used, there is a visible peak of around 450 nm blue light in the white light spectrum, which could suppress melatonin secretion during the night, both at work and at home. Even a smartphone LED may be a source of artificial light at night, which influences the circadian regulation of the sleep–wake cycle, suppresses melatonin secretion, alters mood and cognitive functions, and evokes fatigue. Although the light emitted by LEDs appears white, it has peak emission in the blue light range.

Exposure to high intensity blue light can affect many physiological functions, and can even induce retina injuries or damage. This is why the blue light hazard arising from artificial sources must be evaluated and exposure levels at workers' eye positions must be compared with exposure limit values for blue light hazard established by the International Commission on Non-Ionizing Radiation

Protection (ICNIRP) and Directive 2006/25/EC of the European Parliament and of the Council. However, present-day knowledge about blue light can also be used to treat sleep disorders. In modern society, the use of blue light is becoming increasingly prominent. A large world population is exposed to artificial light at unusual times of the day or late at night. Light has a cumulative effect of many different characteristics (wavelength, intensity, duration of exposure, time of day). It is important to consider the spectral output of light sources for the improvement of alertness during night shifts, but also to minimize the danger associated with blue light and artificial light exposure.

1.5   NEW IDEA OF LIGHTING DESIGN – HUMAN-CENTRIC LIGHTING (HCL)

So far, lighting has often been designed only to ensure safety and appropriate conditions for performing visual work, while maintaining the greatest possible comfort of vision and human well-being. Usually, the non-visual mechanisms of light's impact on the human body were not taken into account. This approach has been changing since the discovery of ipRGCs and the role of the non-visual effects of light. The point-of-view on lighting design changed in the early 2000s, when the innovative idea of dynamic lighting design was born. Moreover, a few years ago, together with the development of knowledge about the impact of light on the human circadian system and new circadian metrics, a new stage began in the design of lighting, focused on human health and well-being, which is called human-centric lighting (HCL) or integrative lighting.

The aim of human-centric lighting is to benefit human health and well-being in various ways resulting both from the effects of light on the visual and non-visual system. Figure 1.1 shows that the same aspects (features) of light, i.e. amount (intensity), spectrum (SPD), spatial distribution, timing, and duration, impact both the visual and non-visual systems, but they have different effects.

The effects of light can be assigned to three basic groups: *visual*, *biological*, and *psychological* (emotional). The *visual* group mainly takes into consideration lighting parameters which influence acuity, visual performance, appearance, and safety. The parameters from this group are supposed to provide high-quality lighting for human vision. The *biological* group mainly looks at circadian/melanopic metrics of light, which influence our circadian performance (phase-shift and alertness). The parameters from this group are supposed to provide high-quality lighting for the human circadian system. The *psychological* (emotional) group is concerned with the users' mood, behavior, and comfort, which are affected both by visual and biological parameters. All three groups together focus on human individual needs, performance, health, well-being, satisfaction, and comfort.

Nevertheless, there is no consensus in the scientific community on the circadian metric which should be taken for designing light. Hence, various metrics are used for this purpose, of which the two most commonly used ones are circadian stimulus (*CS*) and equivalent melanopic lux (*EML*). How to design lighting so as to take into account the non-visual effect of light is also controversial. Moreover, there is a huge problem with what kind of lighting to design for night shift work.

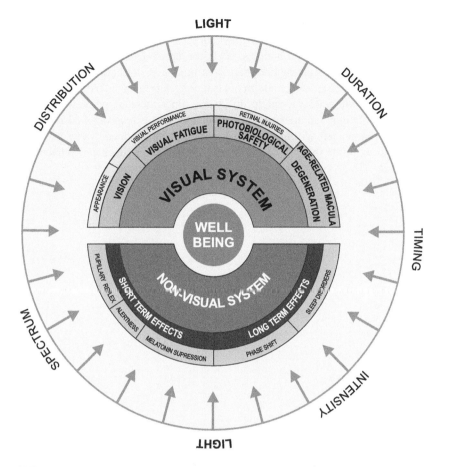

FIGURE 1.1   Visual and non-visual systems are all, in different ways, impacted by lighting.

1.6   WHAT IS THIS BOOK ABOUT?

The book presents a physiological explanation of the visual and non-visual effects of light on humans. It includes two chapters (Chapters 2 and 3), which discuss the biological bases of image- and non-image-forming vision at the cellular level. These may be of particular interest to doctors and medical students, as they provide a thorough overview of the latest discoveries in these fields.

Since the circadian system of shift workers is permanently disturbed, the chapters devoted to the new approach to lighting design at workplaces focus on adopting lighting parameters that could fulfill the needs of both visual and non-visual systems and take into account the development of knowledge in the field of colorimetry and measuring techniques, as well as new circadian metrics for assessing light influence on the non-visual system and regulating human sleep patterns. An in-depth discussion of such lighting design is presented.

Moreover, it is one of the intentions of this book to put forward some recommendations and examples of lighting design which take into account both the visual and non-visual effects of light on humans. Based on up-to-date scientific research and the practical advice of lighting designers presented at various conferences, training courses, or workshops, a compendium of current packages for such projects was presented, also including ideas for night shift work. Since there are no standards for the measurement of circadian metrics, some guidelines in this respect are also included.

We hope that readers will find the book easy to understand, enjoyable, and useful. It is also our ambition that it will guide them in their future decisions related to the use of light.

2 The Biological Bases of Photoreception in the Process of Image Vision

A vision is not just a picture of what could be; it is an appeal to our better selves, a call to become something more.

Rosabeth Moss Kanter

2.1 OPTICS OF THE EYE – FROM THE PUPIL TO THE RETINA

Numerous studies have been devoted to vision and visual perception in human beings. Though far from easy, it is important to understand the complex and still not entirely discovered workings of sight. The following chapter will try to elucidate the biology of photoreception in the process of image vision.

Figure 2.1 provides a diagram showing a schematic cross-section of the human eye (A) and the morphology of the retina (B). Light enters the eye through the cornea (a transparent external surface) and passes through the pupil (the aperture that allows it in). The diameter of the pupil is controlled by the muscles of the iris so that an optimum amount of light can be let in under different conditions. Behind the iris lies the crystalline lens focusing the incoming light on the retina. Thus, the eye can be compared to a photographic camera: it has a variable aperture system (the pupil), a lens system, and a retina that corresponds to the film.

The amount of light entering the eye through the pupil is proportional to its area. The pupil diameter in the human eye can vary from 1.5 mm to 8 mm and is regulated by the iris, whose major function is to admit more light in the dark and less in daylight. The variations in the pupil's aperture make it capable of allowing 30-fold changes in how much light enters the eye.

The lens has just the right curvature for parallel rays of light to pass through each of its parts and be bent exactly enough for all the rays to pass through a single focal point. The more a lens bends the light rays, the greater is its refractive power measured in terms of diopters. In children, this refractive power can be increased from 20 diopters to about 34 diopters, which is an accommodation of about 14 diopters. The elastic lens capsule can change shape (become more or less spherical) in response to the activity of the ciliary muscle, controlled by the autonomic nervous system. This refractive power influences visual acuity, or clarity of vision – in the human eye it is about 25 seconds of arc for discriminating between point sources of light. A person with normal visual acuity looking at two pinpoint light spots 10 m away can barely distinguish the separate spots when they are 1.5 to 2 mm apart.

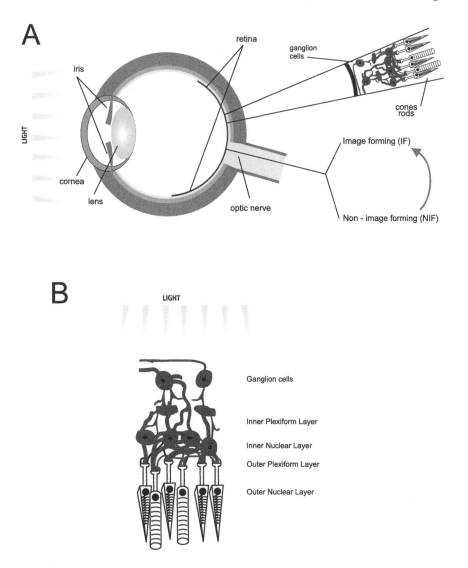

FIGURE 2.1   Cross-section of the human eye (A) and morphology of the retina (B).

The retina, a part of the eye and a specialized part of the central nervous system, called "the brain's window to the world", is the sensory part of the eye [Mayeli 2019]. Its primary job is to convert energy from the sensory stimuli – the photons of light – into an electrical signal, transmitted and analyzed by special regions of the brain in a process called phototransduction. Histologically, the retina consists of three cellular layers, which contain five cell types separated by two synaptic layers. The complex organization suggests the different cells' participation in various physiological regulatory processes. Phototransduction is carried out in light-sensitive neurons called photoreceptors, located at the rear of the retina. The human retina includes two classes of photoreceptors: rods and cones. These make synaptic connections with bipolar cells, which in turn convey information to the retinal ganglion cells, along a "vertical" visual pathway. The axons of the ganglion cells constitute the optic nerve. The signals along this pathway are modulated by inhibitory neurons at two levels: those of the horizontal cells in the outer retina and the amacrine cells in the inner retina [Chapot et al. 2017]. Supporting glial cells (Müller cells) and their cytoplasmatic processes fill the space between photoreceptors and bipolar and ganglion cells. Müller cells are essential for the transmission of light, due to their unique shape, orientation, and refractive index [Franze et al. 2007]. They act as conduits that enable light to reach the photoreceptors with minimal scattering. It seems that every Müller cell is coupled with a partner cone cell. In addition to these cells, microglial cells are present in all layers.

2.2 PHOTORECEPTORS

There are two types of photoreceptors: rod cells and cone cells, which number about 120 million and 6 million in each eye, respectively, forming an average input of 100 photoreceptors to each of the 1.5 million retinal ganglion cells [Molday et al. 2015; Bloch et al. 2019]. Rod photoreceptors are highly sensitive and responsible for low-light vision. Only seven rods are needed to detect a single photon of light in a dark environment [Nicholls et al. 2001; Jerath et al. 2016]. Cone photoreceptors are responsible for daylight and color vision, because they respond to a broad range of light intensities of specific wavelengths [Ueno et al. 2018]. Cones are less sensitive to light and cannot detect it if there are fewer than 100 photons [Jerath 2016]. The fovea consists of a concentrated number of elongated cones for color vision acuity in bright light, and extra-foveal regions for acuity in dimmer light [Jerath 2016]. Morphologically all photoreceptors consist of two parts: the outer segment – the light-sensitive portion of the photoreceptor – and the inner segment, containing the nucleus and the usual cellular make-up. Their synaptic terminal makes synaptic connections with the bipolar and horizontal neurons of the retina. The outer segment is long and cylindrical in rods, whereas in cones it is shorter and typically tapers toward the distal end. The membrane resting potential of a photoreceptor cell in the dark has a value of -30 to -40 mV. If the light is turned on, the membrane potential shifts to near -70 mV. Thus, the response to light, called the receptor potential of the photoreceptor, is a hyperpolarization. The degree of hyperpolarization depends on the light intensity. A light stimulus hyperpolarizes photoreceptors in the outer segment and reduces the release of neurotransmitters by decreasing a Ca^{2+} influx at their synaptic terminals. Nonselective cation channels in the cell membrane of the outer segment are open in the dark, maintaining the membrane in a depolarized state. Light closes these channels. The fraction of closed channels depends on the intensity of light, which explains the different changes of the membrane potential observed in response to bright and dim light stimuli. The light-absorbing molecules – the visual pigment molecules – are involved in the process of detecting the presence of light. The outer segments of rods and cones are arranged in multiple layers, with membranes containing pigment molecules. A photon of light passes through a series of these, each with a high density of light-absorbing molecules. Thus, photons reaching the outer segments are absorbed and detected by photoreceptors. The human retina has four types of photoreceptors: three for cones and one for rods, with three kinds of visual pigments (photopsins) in the cones and rhodopsin in the rods. Each class absorbs light in the visible part of the electromagnetic spectrum at 400 nm to 700 nm, but is maximally sensitive to light in its specific wavelength. Cone opsin pigments have been less well-described, due to their lower availability and their higher conformation instability [Srinivasan et al. 2015]. The rod photoreceptor cell system and its visual pigment, rhodopsin, has been much more extensively studied. Rhodopsin contains a large protein molecule (opsin) with a chromophore, which absorbs light. The chromophore is the same for all types of visual pigments, i.e. retinal – the aldehyde of vitamin A1. Cone opsin pigments belong to the superfamily of G-protein-coupled receptors consisting of a transmembrane apoprotein and retinal. In an inactive dark-adapted state, the retinal chromophore is covalently bound to the visual pigments through its aldehyde group via a protonated Schiff base linkage (PSB). The basis of the sense of sight is a process of photoisomerization of retinal: in the dark, retinal is in the 11-*cis* conformation, but after absorbing a photon of light, it undergoes a conformational change to the all-*trans* isomer. The primary event of the visual transduction cascade after retinal isomerization is deprotonation and breakage of PSB linkage, and retinal release from the retinal binding pocket. Thus, this change of conformation induces conformational changes in the opsin protein, which in interaction with other proteins in the outer segment results in the closing of cation channels and in the hyperpolarizing light response of the photoreceptors. This response is generated by a cGMP signaling cascade, which leads to cGMP hydrolysis and to the closure of cGMP-gated, nonselective cation channels open in darkness [Finn et al. 1998]. For example: during phototransduction, photoisomerized rhodopsin activates G-protein transducin, the transducin alpha subunit targets the enzyme phosphodiesterase (inactive in rods when it is dark); the activated phosphodiesterase inactivates cyclic GMP (the second messenger in cells); the reduction in cyclic GMP concentration causes the nonspecific (nonselective) cation channels to close, and the photoreceptor hyperpolarizes.

Photovoltage is shaped by voltage-gated channels in the inner segment (L-type Ca^{2+} current, a delayed rectifier K^+ current, a fast, transient K^+ current, and a hyperpolarization-activated cation current).

Jerath et al. [2016] suggest that photoreceptors generate the dark current from photopigments in the disks of photoreceptors. Dark currents exist without illumination and increase when there is light.

As described above, highly sensitive rod photoreceptors are responsible for low-light vision, whereas cone photoreceptors are less sensitive, but respond to a broad range of light wavelengths, and thereby they are responsible for daylight and color vision. Over 100 mutations identified in rhodopsin are involved in various ocular impairments, including congenital stationary night blindness and retinitis pigmentosa [Ortega et al. 2019; Athanasiou et al. 2018]. There is also great interest in finding novel ligand molecules that would improve the folding and stability of rhodopsin mutants [Ortega et al. 2019; Behnen et al. 2018]. Ortega et al. [2019] suggested that flavonoids could be utilized as lead compounds in the development of effective non-retinoid therapeutics for managing retinitis, i.e. pigmentosa-related retinopathies.

2.3 CELLULAR ORGANIZATION OF THE VISUAL PATHWAYS

The schematic organization of connections between cells in the retina involved in the visual pathway is well documented:

- photoreceptors transmit signals to the outer plexiform layer, forming synapses with bipolar and horizontal cells;
- horizontal cells transmit signals horizontally from the rods and cones to bipolar cells;
- bipolar cells transmit signals vertically from the rods, cones, and horizontal cells to the ganglion and amacrine cells;
- amacrine cells transmit signals directly from bipolar cells to ganglion cells, or horizontally from the axons of bipolar cells to dendrites of the ganglion cells or to other amacrine cells;
- ganglion cells transmit output signals from the retina through the optic nerve to the brain centers.

2.3.1 SYNAPTIC CONNECTIONS – PHOTORECEPTOR CELLS – BIPOLAR CELLS

The human retina processes image-forming visual information from cone and rod photoreceptors through parallel ON and OFF pathways [Schiller 1992; Schiller 2010]. The photoreceptors make synaptic connections with two types of bipolar cells, transmitting signals into parallel ON- and/or OFF-bipolar cells. The presynaptic terminals of photoreceptors release the neurotransmitter glutamate. The rate of glutamate release is high in darkness (depolarization of the photoreceptors) and reduced in response to light (hyperpolarization of the photoreceptors). The first step is the segregation of the cone signals to the ON and OFF pathways: the cone information diverges to ON- and OFF-bipolar cells. Thus, both types of cone bipolar cells (ON and OFF) receive inputs from cone photoreceptors and transmit directly to retinal ganglion cell dendrites in the inner plexiform layer (see Figure 2.1B). There are only ON-bipolar rod cells whose axon terminals extend to the deeper region of the inner layer. Thereby, rod photoreceptors transmit signals only through the ON pathway. Dendrites of ON-bipolar cells express the metabotropic glutamate receptor 6 (mGluR6, Grm6). The ON-bipolar dendrite has a single invaginating contact with a photoreceptor presynaptic axon terminal [D'Orazi et al. 2014; Ueno et al. 2018]. The OFF-bipolar cells present inotropic GluRs (AMPA/kainate receptors), glutamate-gated cation channels on their dendrites. The OFF-bipolar dendrite forms multiple flat contacts with a cone photoreceptor presynaptic axon terminal [D'Orazi et al. 2014; Ueno et al. 2018]. In the dark, glutamate released from cone presynaptic axon terminals *depolarizes OFF-bipolar cells*, acting through GluRs and *hyperpolarizes ON-bipolar cells*, acting through mGluR6, the activation of which leads to the closure of TRPM1 channels (transient receptor potential cation channel subfamily M member1). The TRPM1 channel in ON-bipolar cells is gated by both the α and the β gamma subunits of the G-protein Go [Xu et al. 2016]. Consistently, OFF-bipolar cells become more negative in the presence of light (inhibitory hyperpolarizing signal), whereas ON-bipolar cells show a depolarizing light response (excitatory depolarizing signal). The importance of the phenomenon of the two types of bipolar responses lies in the fact that it provides a mechanism for lateral inhibition. Depolarized and hyperpolarized bipolar cells lie immediately against each other, which provides a mechanism for separating contrast borders in the visual image. Thus, lateral inhibition is considered to be the first stage in vision processing.

Recently, 15 types of bipolar cells were defined in the mouse retina owing to using several methods [Shekhar et al. 2016; Tsukamoto et al. 2017]. Morphological differences among the types are subtle. Pang et al. [2010] suggested two distinct groups of rod bipolar cells: RB1 has a deeper axon terminal and fewer chloride channels as compared to the RB2 group. Functionally, three types of bipolar cells seem to play a crucial role in the process of image-vision:

- one of these connects only to rods (a type of visual system present in many lower animals) with output only to the amacrine cell, which relays the signals to the ganglion cells. The pure rod vision pathway involves four neurons: a rod, a bipolar cell, an amacrine cell, and a ganglion cell;
- the other two bipolar cells are connected with rods and cones, with direct outputs to ganglion cells by way of amacrine cells. The signal from a single rod is represented as more than 100 replicates at the axon terminal ribbon synapses of two rod bipolar cells, which are directed mainly toward two or three rod-driven amacrine cells (type AII, see Section 2.3.5) [Tsukamoto et al. 2013, 2017]. Probably all bipolar cell types have chemical or electric synaptic contacts with AII amacrine cells.

2.3.2   Synaptic Connections – Bipolar Cells – Horizontal Cells (Ribbon Synapse)

Photoreceptor cells form a specialized synapse with bipolar cells and horizontal cells – the *ribbon synapse*. Cone photoreceptor axon terminals contain many synaptic sites, each represented by a presynaptic structure called a ribbon. Usually rods feature only a single ribbon, whereas cone terminals have multiple ribbons. Each ribbon relays the light signal to one ON cone bipolar cell and several OFF cone bipolar cells [Chapot et al. 2017]. The ribbon synapse contains invaginations, including bipolar dendritic tips and horizontal cell processes. The formation of photoreceptor synaptic invagination requires a composition of an extracellular matrix protein, pikachurin (localized in the synaptic cleft between photoreceptors and ON-bipolar cells), and a membrane glycoprotein, dystroglycan (localized in photoreceptor synaptic terminals) [Kanagawa et al. 2010; Omori et al. 2012]. The loss of these factors causes the absence of the proper intrusion of ON-bipolar cell dendritic tips into photoreceptor axon terminals, resulting in the impairment of synaptic signal transmission between photoreceptor and ON-bipolar cells. The transmembrane protein ELFN1, involved in synapse formation and synaptic function, is expressed in rod photoreceptors and participates in synapse formation between rod photoreceptors and rod ON-bipolar cells [Tomioka et al. 2014; Cao et al. 2015]. Ueno et al. [2018] found that "Lrit1, a leucine-rich transmembrane protein, localizes to the photoreceptor synaptic terminal and regulates the synaptic connection between cone photoreceptors and cone bipolar cells". The results obtained by these researchers suggest that the Frmpd2-Lrit1-mGluR6 axis regulates selective synapse formation in the cone photoreceptor and is essential for normal visual function.

2.3.3 SYNAPTIC CONNECTIONS – HORIZONTAL CELLS

It is also worth discussing the next synaptic connection between photoreceptors and horizontal cells – multi-purpose interneurons of the outer retina. The receptive field of each horizontal cell is very large, larger than its dendritic field [Shelley et al. 2006]. The dendritic arbor and the axon terminal system make a network through the Connexin 57-formed gap-junction [Hombach et al. 2004]. Horizontal cells form an electrically coupled lateral network. When light falls on the photoreceptors, making synaptic connections with some horizontal cells, evoked hyperpolarization spreads laterally via the electrical synapses (electrical junctions) into neighboring horizontal cells. The receptive field of a horizontal cell is larger if compared to the activation from a directly connected group of photoreceptors.

Horizontal cells, like photoreceptors, are depolarized in darkness – when glutamate is released by the photoreceptors – and hyperpolarized in the light. Cones release glutamate, binding to postsynaptic receptors on horizontal (inotropic AMPA- and kainate-type glutamate receptors) and bipolar cell dendrites [Chapot et al. 2017]. Horizontal cells form synapses with the cones (feed back onto cones) and modulate their glutamate release [Feigenspan et al. 2015]. Three different feedback mechanisms have been studied: ephaptic, proton (pH)-mediated, and GABA-mediated. Voltage-gated Ca^{2+} channels, located at the active zone of the ribbon synapse in the cone axon terminal are effectors of the ephaptic and proton-mediated feedback process, regulating the cone glutamate release.

2.3.4 SYNAPTIC CONNECTIONS – THE CONE SYNAPSE

Thus, a cone synapse is formed by three different neurons: photoreceptor cells, horizontal cells, and bipolar cells. This synapse, which plays a fundamental role in the visual system, encodes information from the outside world. It is very complex and still elusive. The first process of *lateral inhibition*, averaging the input from many cones and subtracting this mean signal from the local cone response, might be involved in visual contrast enhancement. The following explanation of the process of lateral inhibition seems to be most correct: the outputs of horizontal cells are inhibitory; the visual pathway from the central area, where the light strikes, is excited, and an area to the side is inhibited; and transmission through the horizontal cells puts a stop to the signal widely spreading in the retina.

A recent study indicates the structure of this interaction is more complex: individual cones form feedback at the level of a single horizontal cell dendrite. Additionally, it seems that primates possess two independent horizontal cell networks (divisions): HI cells shaping two classes of light-sensitive cones: M- and L-cone output, and HII cells shaping another class: S-cone output [Chapot et al. 2017].

2.3.5   SYNAPTIC CONNECTIONS – THE ROLE OF AMACRINE CELLS

Amacrine cells are interneurons, involved in the analysis of visual signals before they leave the retina, mediating the lateral interconnections. They are lateral inhibitory interneurons which produce effects at the level of bipolar cell output synapses (reciprocal synapses onto bipolar cell synaptic terminals). Glycinergic amacrine cells are the narrow-field type (AII amacrine cells), and GABA-ergic amacrine cells are the wide-field type [Tsukamoto et al. 2017]. Both dendrites and the axon-like-processes of dopaminergic amacrine cells narrowly stratify close to the outer border of the inner plexiform layer, forming the "off" switch for the AII-mediated rod pathway. Morphological and physiological studies make it possible to identify about 30 types of amacrine cells. One of these is part of the pathway for rod vision (AII) briefly described above: from rod to bipolar cells to amacrine cells to ganglion cells. Another type responds strongly at the onset of a continuing visual signal, while still another one at the offset of a visual signal. Both responses fade quickly. Other amacrine cells, which are directionally sensitive, respond to the movement of a spot across the retina in a specific direction. A strong correlation was found between the number of amacrine cell synaptic contacts and the number of bipolar cell axonal ribbons.

Formation of bipolar cell output at each ribbon synapse may be effectively regulated by a few nearby inhibitory inputs of amacrine cells. All the amacrine cells are connected to five of the six OFF bipolar cell types via chemical synapses and seven of the eight ON cone bipolar cell types via electrical synapses (gap junctions).

[Tsukamoto et al. 2017]

Depolarization of light responses of all types of amacrine cells is transient, producing a depolarization only briefly at the onset or termination of light, or at both the onset and termination of illumination. ON-type amacrine cells receive inputs only from ON-bipolar cells, and depolarize at the onset of illumination, and OFF-type amacrine cells receive synaptic inputs from only OFF-bipolar cells, and depolarize at the end of illumination. ON-OFF amacrine cells receive synaptic inputs from both ON- and OFF-bipolar cells. Amacrine cells also fire action potentials when the depolarizing synaptic response is sufficiently large. This is the first action potential considered as a light response of the retinal neurons (in the photoreceptors, horizontal cells, and bipolar cells, the changes in membrane potential in response to illumination spread by electronic conduction).

Despite a lot of data, the nature of the complex interconnections between retinal neurons in the process of light signal transmission to ganglion cells still remains uncertain.

2.3.6   SYNAPTIC CONNECTIONS – FROM RODS AND CONES TO GANGLION CELLS

The output signal from the retina is transmitted to the brain's visual center via the axons of the retinal ganglion cells, which form the optic nerve. The electrical signal used by ganglion cells is the action potential, because the information is carried over long distances. Information coming from about 100 million rods and 3 million cones (as described previously) converges on about 1.6 million ganglion cells: 60 rods and 2 cones converge on each ganglion cell. In the central part of the retina the rods and cones become slenderer, while in the central fovea there are only cones and no rods. This fact explains the higher visual activity in the central retina as compared to the poorer acuity peripherally. However, the peripheral retina is more sensitive to weak light. Its greater sensitivity could be explained by the fact that about 200 rods converge on a single fiber of optic nerve, giving summarized signals from the rods. The sizes of receptive fields of ganglion cells vary, commonly having a concentric center-surround organization. Light in the center of the receptive field of an ON-type ganglion cell increases the frequency of their action potential firing, while light in the surrounding regions inhibits the cell. Light in the center of the receptive field of an OFF-type ganglion cell inhibits the cell, while light in the surrounding regions excites the cell. The center-surround organization of the ganglion receptive fields produces some visual illusions; the visual system perceives differences in light intensity when no such difference is actually present in the stimulus. Histologically, two classes of ganglion cells exist: diffuse ganglion cells, contacting several bipolar cells, and midget ganglion cells, whose dendrites contact a single midget bipolar cell, receiving inputs from cones only. One class of retinal ganglion cells – alpha cells – have large cell bodies and widely branching dendritic fields. These cells are found predominantly in the peripheral retina, receiving inputs mainly from rods. They respond to rapid changes in visual images when new visual events occur anywhere in the visual field, but they do not specify the location of this event with great accuracy. These cells are also called M cells, because in humans they connect to large cells in the magnocellular layers in the lateral geniculate nucleus. A second class of retinal ganglion cells – beta cells – with medium-sized cell bodies, have more restricted dendritic fields. These cells are found predominantly in the central retina, receiving inputs mainly from cones. Their signals represent discrete retinal locations and transmit fine details of visual images. Beta cell transmission is responsible for color vision. (They show a center-surround organization, where the center responds to one color, and the surround maximally responds to the color opposite in a color wheel. These cells are also called P cells, because in humans they connect to smaller cells in the parvocellular layers in the lateral geniculate nucleus). Other ganglion cells are histologically classified as gamma, delta, and epsilon cells. Physiologically, they transmit signals at slow velocity, receiving most of them from rods, transmitted by way of bipolar cells and amacrine cells. They are responsible for crude rod vision under dark conditions. There are some physiological differences between M and P cells: the receptive fields for P cells are smaller than for M cells; P cell axons conduct impulses more slowly than M-cell axons; and the responses of P cells, especially to color stimuli, can be sustained, whereas the responses of M cells are more transient. While P cells are generally sensitive to color, M cells are not sensitive to color stimuli, but they are much more sensitive than P cells to low-contrast as well as black-and-white stimuli.

Retinal ganglion cells represent the last stage of retinal processing and the first stage in providing responses in the form of action potentials along the visual pathways. They share properties with cortical neurons, such as anatomical structure, functions, and neurotransmitters (dopamine, serotonin, glutamate, and GABA)

[Schwitzer et al. 2017a; Bernardin et al. 2019]. Retinal ganglion cell function has been studied in various psychiatric disorders (using electroretinogram recordings): major depressive disorder [Bubl et al. 2010], autism [Tebartz-van Elst et al. 2015], attention deficit hyperactivity disorder [Bubl et al. 2015], cannabis use disorder [Schwitzer et al. 2017b; Schwitzer et al. 2018], and in schizophrenia [Demmin 2018]. Recently, Bernardin et al. [2019] demonstrated a slowing of retinal ganglion cells signaling in schizophrenia patients with visual hallucinations, which could affect the quality of visual information reaching the visual cortex. The data collected suggest that changes of the physiological function of the ganglion cells might be involved in the pathogenesis of psychiatric disorders. Wostyn [2020] in a study from this year suggests that retinal nerve fiber layer (RNFL) thinning may occur in patients with Chronic Fatigue Syndrome (CFS) and may serve as "an ocular biomarker of underlying glymphatic system dysfunction" in CFS and in neurodegenerative diseases that result from protein toxicity [Wostyn 2020].

2.4 CALCULATION OF COLOR AT THE LEVEL OF GANGLION CELLS

In low light levels (luminance <0.001 cd/m^2), below the cone threshold, vision is *scotopic*: visual response relies entirely on rod signals (luminance is a photometric measure of the luminous intensity per unit area of light traveling in a given direction). Scotopic vision is characterized by poor visual acuity and lack of color discrimination. In high light levels (luminance ≥ 3 cd/m^2), above rod saturation, vision is *photopic*: visual response relies entirely on cones. Photopic vision is characterized by good visual acuity and color discrimination. In the intermediate light levels (luminance between 0.001 and 3 cd/m^2) between cone threshold and rod saturation, both rods and cones contribute to visual response. This kind of vision, when both rods and cones provide signals to the retinal ganglion cells, is defined as *mesopic* vision.

Human color vision is mediated by three classes of light-sensitive cones with peak sensitivities at the short (S, 420 nm), medium (M, 530 nm), and long (L, 560 nm) wavelengths, active under photopic levels of illumination. The cones sensitive to short-wavelength light have a photopigment, absorbing light best in the violet and blue part of the visual spectrum. The cones sensitive to middle-wavelength light have a photopigment best absorbing light in the green and yellow part of the spectrum, whereas the cones sensitive to long-wavelength light have a pigment best absorbing light in the yellow and red portion of the spectrum. The characteristic differences in visual sensitivities are due to specific amino acid variations in the retinal binding pocket, which result in different absorption maxima as a result of the so-called opsin shift effect [Srinivasan et al. 2015; Merbs et al. 1992]. Rods have peak sensitivity at about 498 nm – they are active at lower levels of illumination (scotopic). Color vision results from two opponent processes: from the evolutionarily older yellow–blue opponent system (dichromatic vision) and the red–green opponent system, which evolved much later (trichromatic vision). Most humans are trichromatic, a small number of people are dichromatic (with color-defective vision), and some females are tetrachromatic [Westland 2017; Mollon et al. 2001; Jordan et al. 1993]. In color opponency the signals from two different photoreceptor classes with different spectral tuning are subtracted, thereby encoding changes in the relative wavelength content of the light that occur during dusk and dawn. [Spitschan et al. 2017c]. Recent years have provided new data concerning the role of photopigments. The findings of Srinivasan et al. [2015] have for the first time revealed differences in the photoactivated conformation between red and green cone pigments and support the proposal of secondary retinal binding to visual pigments, additional to binding to the canonical primary site, "which may serve as a regulatory mechanism of dark adaptation in the phototransduction process" [Srinivasan et al. 2015].

Humans cannot perceive color when the level of illumination is so dim that only rod photoreceptors are active (moonlight); this type of photoreceptor has a single type of photopigment. The light input represents a mixture of different wavelengths to the eyes. The visual system should and can determine the contribution of light from different wavelengths by analyzing neural output from each type of cone photoreceptor. This comparison occurs at the level of color-sensitive ganglion cells. Some types of ganglion cells respond best to a specific color of light (the suitable wavelength). Color-sensitive ganglion cells have center-surround receptive fields, and the center and surround responses originate from different cones. Illumination in the center produces the opposite effect to illumination in the surround region. Thus, the response depends on the wavelength of light entering the central and surround regions of the receptive field. For example, the excitatory receptive center can be mediated via L-cones (more sensitive to red light), and the inhibitory surround region via M-cones (more sensitive to green light). If a red light covers the receptive field of this neuron, the cell will be excited. If a green light covers the receptive field, the M-cones of the inhibitory surround will be stimulated, resulting in the inhibition of the ganglion cell (excitation by red light and inhibition by green light). The converse organization – the center originating from M-cones and the surround from L-cones – produces a converse response: excitation by green light and inhibition by red light. Ganglion cells demonstrating opposing responses to red and green lights are called red–green opponent ganglion cells (Figure 2.2). Other ganglion cells, producing opposing responses to blue lights (affecting S-cones) and yellow lights (affecting M-cones), are called blue–yellow opponent ganglion cells (Figure 2.3). The center and the surround of ganglion cells derive from a single type of cone photoreceptor. Color-sensitive cells of this type are called single-opponent cells.

The perception of white is formed by mixing colors of just a few wavelengths, such as red, green, and blue, or just a pair of complementary colors, such as blue and yellow.

2.5   HIGHER VISUAL PROCESSING

The axons of the retinal ganglion cells form two optic nerves, which further, at the optic chiasm, divide axons in two groups:

- axons coming from ganglion cells in the nasal half of the retina cross to the opposite side;
- axons coming from ganglion cells in the temporal half of the retina remain on the same side.

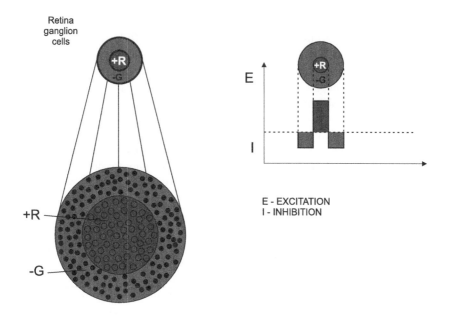

FIGURE 2.2   Opposing ganglion cells responses to red and green lights: red–green oppo-nent ganglion cells.

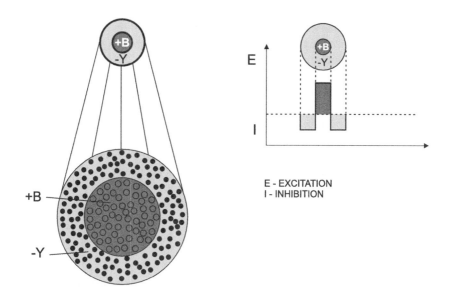

FIGURE 2.3   Opposing ganglion cells responses to blue and yellow lights: blue–yellow opponent ganglion cells

Thus, the nasal retina from the left eye and the temporal retina from the right eye (the left part of the visual field) are represented in the right half of the brain, and the right part of the visual field in the left half of the brain.

In this way, visual information is transmitted to the lateral geniculate nucleus of the thalamus (a gate controlling passage of information to the primary visual cortex in the occipital lobe). In this nucleus, color-sensitive projection neurons have the same receptive field organization as those found in ganglion cells. At the level of the cortex, color-sensitive neurons and the organization of their receptive fields differ from the previously described mechanisms. Color opponency appears within the center and surround of the receptive field and also between the center and surround (*double-opponent color-sensitive cells*). Some cortical neurons are orientation-selective and motion-sensitive. The primary visual cortex carries out an analysis of the visual field for a variety of aspects of visual stimuli, such as orientation, direction of movement, and object color [Matthews 1998]. Further, visual information is transmitted to higher-order cortical areas – anterior to the primary visual cortex – for the processing of specialized information. Such functions as the processing of object form, color, or motion project to more specialized areas in a more anterior part of the occipital lobe and the posterior part of the temporal lobe cortex [Matthews 1998].

Recently, Jerath et al. [2016] presented a new theory of vision: retinal cells (photoreceptors, horizontal, bipolar, amacrine, ganglion, and Müller cells) are oscillating at alpha rhythms and are synchronized with visual and parietal cortices. "This creates a visual space that forms an infrastructure needed for vision when photoreceptors are stimulated. The cortex sends feedback to the layers in the visual pathway in the retina and lateral geniculate nucleus in the thalamus". Such ideas suggest that the brain and retina communicate in real time. Furthermore, these authors consider that retino-geniculo-cortical synchronization results in the final electrical image observed from the lateral geniculate nucleus, which – as they postulate – "could also be from the fovea, because we observe our world internally from the thalamus".

2.6   THE ROLE OF PURINERGIC SIGNALING IN THE RETINA

Purinergic signaling regulates several events in the developing retina and in adult damaged tissue [Ventura et al. 2019]. Adenosine is involved mainly in cell survival. Nucleotides induce the proliferation and migration of retinal cells, through activation of P2Y receptors. On the other hand, activation of the Pr2X7 receptor subtype can induce cell death in the developing and damaged retina. It seems that activation of nucleotide receptors might be involved in the self-repair process of the retina by inducing pluripotency genes in glial cells. This hypothesis is very attractive for drug-based strategies to prevent and/or treat retinal diseases [Ventura et al. 2019].

2.7   AGE-RELATED CHANGES IN THE OPTICS AND RETINA OF THE EYE

Changes occurring in the eye and its connections with the brain with age have several effects on vision, reducing visual performance and comfort.

2.7.1   LOSS OF LENS ELASTICITY – LOSS OF ACCOMMODATION

In the elderly, the lens becomes less elastic, because of progressive denaturation of the lens proteins. The power of accommodation decreases from 14 diopters to less than 2 diopters in individuals who are 45–50 years old, and decreases to 0 diopters at the age of 70 (which means that the lens remains totally non-accommodating).

2.7.2   LENS YELLOWING – CHANGES OF PERCEIVED COLORS

The yellowing process of the eye lens starts at the age of 25 and is most pronounced in people over 75 years old. This effect is called the "blue light loss effect", because blue light is largely absorbed by a yellow lens.

2.7.3   DAMAGE OF PHOTORECEPTOR AND GANGLION CELLS

Damage or loss of some photoreceptor and ganglion cells reduces the visual signal reaching the visual cortex. With growing age (especially over the age of 60) a reduction of photopigment is observed, resulting in loss of the conversion of light into a neural signal.

2.7.4   STRUCTURAL CHANGES IN THE OPTICS

Structural changes of the cornea, lens, eyeball (clouding), retina, and blind spot result in an increase in glare. They all have negative consequences for the amount of light reaching the retina.

2.7.5   DISTURBED ROD PIGMENT REGENERATION

Impaired regeneration of rod photopigment – regeneration of rhodopsin – takes more time. It results in slower adaptation to the dark, reducing visual performance and less comfort in the dark.

2.8   SUMMARY

Phototransduction – the process of the translation of photon energy into electrical signals – occurs in photoreceptors located in the retina. Rods (one class of photoreceptors) are active in dim light, whereas cones (another class of photoreceptors) function in brighter light. Photons of light are absorbed by the visual pigments of molecules in the photoreceptor cells. The pigment molecule is formed by retinal (vitamin A1 aldehyde) and a large membrane protein, called opsin (rhodopsin in rods). Absorption of light causes biochemical and electrophysiological processes in the cells of the retina. The complex organization of the retina suggests the different cells' participation in various physiological regulatory processes in the visual system. Visual information is carried from the retina to the other structures of the central nervous system by patterns of action potentials in the axons of the retinal ganglion cells forming the optic nerve. This group of ganglion cells are the important part of image vision. "The visual brain" is specialized in analyzing object form, color, location, and motion.

3 The Biological Basis of Non-Image-Forming Vision

Vision is the art of seeing what is invisible to others.

Jonathan Swift

3.1 INTRODUCTION

It had long been assumed that light signals for image and non-image vision began with the rods and cones of the retina. It was only the last two decades that have witnessed the discovery of a small population of retinal ganglion cells that express a new photopigment – melanopsin – and are themselves atypical photoreceptors. The work done by Berson et al. [2002] in 2002 conclusively identified the elusive third class of retinal photoreceptors. These cells are intrinsically photosensitive and drive a variety of non-image visual signals.

Intrinsically photosensitive retinal ganglion cells (ipRGCs) are the earliest photoreceptors in mammalian development. Experimental studies showed that ipRGCs in mice are already photosensitive at birth. These neonatal cells are functionally connected to the suprachiasmatic nucleus (SCN), participating in neonatal photoentrainment (whereas rods and cones become photosensitive postnatally) [Do et al. 2010]. In contrast to rhodopsin, melanopsin is detectable in the second week of embryonic development. The sensitivity of ipRGCs increases after birth, becoming more sustained in the first weeks. This represents a maturation of the intrinsic photosensitivity of these cells. The size and complexity of dendritic arbors increase from a diameter of 100 μm around birth to 300 μm by the third postnatal week.

3.2 THE DISCOVERY, FUNCTION, AND DIVERSITY OF INTRINSICALLY PHOTOSENSITIVE RETINAL GANGLION CELLS (ipRGCS); SOME INTERACTIONS WITH THE CONVENTIONAL PHOTORECEPTORS

Melanopsin-containing ipRGCs show typical morphological characteristics. The expression of melanopsin in the soma and the dendrites of iPRGCs distinguishes these cells from other retinal ganglion cells involved in image vision [Do et al. 2010; Belenky et al. 2003; Berson et al. 2002; Do et al. 2009; Provencio et al. 2002]. The dendrites of iPRGCs are irregular and rich in varicosities. The axons of iPRGCs also express melanopsin and are unmyelinated, unlike those in conventional retinal ganglion cells [Do et al. 2010]. Initially, ipRGCs were considered to be a uniform population of all the retinal ganglion cells that can detect light levels (irradiance detectors), involved in circadian photoentrainment (the projection to the suprachiasmatic nucleus – SCN) and in the pupillary light reflex (the projection to the olivary pretectal nucleus – OPN). Subsequent studies have uncovered that ipRGCs consist of several subtypes that are morphologically and physiologically distinct, and in different ways contribute to non-image- and image-forming behavior. The ipRGC family has up to five distinct cell types (M1 to M5), as shown in Figure 3.1 [Schmidt et al. 2011a; Sonoda and Schmidt 2016]. The subclass of ipRGCs that was discovered first is M1 cells. Most of their somata are located in the ganglion cell layer (GCL), although some are displaced to the inner nuclear layer (INL). Their dendritic fields overlap to cover the entire retina, forming a photoreceptive net [Do et al. 2010; Hattar et al. 2002; Provencio et al. 2002]. The dendrites of M1 cells stratify in the inner plexiform layer (IPL) of the retina at the border with the INL, in the so-called OFF-sublamina of the IPL, where OFF-bipolar cells, depolarizing in response to light decrement, arborize their axon terminals [Do et al. 2010]. However, M1 cells receive synaptic input from ON-bipolar cells (depolarizing in response to light increment) [Dumitrescu et al. 2009; Hoshi et al. 2009]. M1 cells show the highest intrinsic photosensitivity of all the ipRGCs subclasses, probably because they have the highest melanopsin immunoreactivity and melanopsin density [Schmidt and Kofuji 2010; Do et al. 2009].

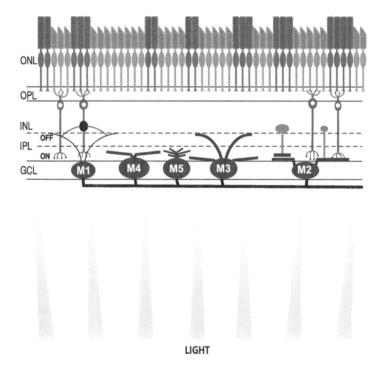

FIGURE 3.1 Location of the five morphological subtypes: M1–M5 of ipRGCs in retina. ONL – outer nucleus layer; OPL – outer plexiform layer.

M2 cells are as numerous as M1 cells and cover the entire retina [Do et al. 2010; Viney et al. 2007]. The dendrites of M2 cells stratify on the opposite side of the IPL from the M1 cells, to the ON-sublamina of the IPL, at the border with the GCL. M2 cells are less intrinsically photosensitive than M1 cells, but they can fire action potentials at higher frequencies than M1 [Schmidt and Kofuji 2010].

The next ipRGC subtype is M3 cells. They stratify their dendrites in both the ON- and OFF-sublaminae of the IPL, showing variability in the proportion of the dendritic stratification in the inner and outer sublaminae [Schmidt et al. 2011c].

A next discovery – two additional subtypes – are the M4 and M5 cells [Ecker et al. 2010]. Both of these cell types stratify in the inner sublamina of the IPL. M4 cells have the largest soma of any described ipRGC subtype and more complex dendritic arbors than M2 cells, while M5 cells have small, highly branched arbors arrayed uniformly around the soma: "bushy dendritic arbors" [Schmidt et al. 2011a; Ecker et al. 2010]. Both of the latter cell types showed a weak response to light, which suggests that M4 and M5 cells express functional photopigment. The diversity within ipRGCs assumes dendritic morphology and stratification, membrane properties, and melanopsin expression.

Atypical ipRGC photoreceptors also act as conventional retinal ganglion cells. They receive synaptic (extrinsic) input from rod/cone photoreceptors via bipolar cells [Schmidt et al. 2011b; Viney et al. 2007]. As described earlier, RGCs with dendrites stratifying in the outer sublamina of the IPL receive excitatory synaptic inputs from OFF-bipolar cells (responding to light decrements). RGCs with dendrites stratifying in the inner sublamina of IPL receive synaptic inputs from ON-bipolar cells (responding to light increments) [Schmidt and Kofuji 2010], while ipRGC subtypes receive inputs respective to their stratification patterns. They receive synaptic inputs according to the scheme: M1 – from OFF-bipolar cells; M2, M4, and M5 from ON-bipolar cells; M3 from both ON- and OFF-bipolar cells. However, contrary to this expectation, M1 (OFF stratifying), M2 (ON stratifying), and M3 (ON-OFF stratifying) ipRGCs receive predominantly ON input [Pickard et al. 2009; Pickard and Sollars 2012; Schmidt and Kofuji 2010]. A weak OFF input to M1 cells has been reported only under pharmacological blockade of amacrine cell inputs [Wong et al. 2007]. The interaction between M1 ipRGCs and amacrine cells remains to be studied in the future. Thus, ipRGCs combine direct photosensitivity through melanopsin with a synaptically mediated drive from classical photoreceptors through bipolar cell input.

3.3 BRAIN REGIONS INNERVATED BY ipRGCS

In mammals, ipRGCs modulate a multiplicity of behaviors: circadian photoentrainment, pupillary light reflex, activity masking, sleep/arousal and neuroendocrine systems, anxiety, and light aversion (Figure 3.2).

The M1 subtype of ipRGCs project to many regions in the brain, showing their functional involvement:

- in the suprachiasmatic nucleus (SCN), the master circadian clock,
- in the intergeniculate leaflet (IGL), a center for circadian entrainment, integration of photic and nonphotic circadian cues,

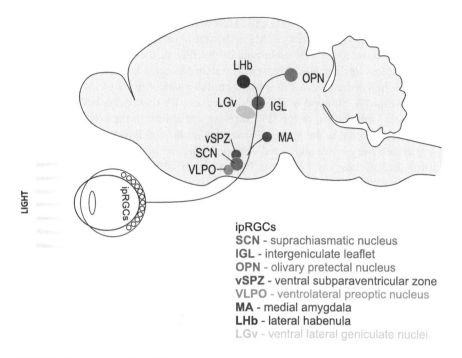

LIGHT

ipRGCs
SCN - suprachiasmatic nucleus
IGL - intergeniculate leaflet
OPN - olivary pretectal nucleus
vSPZ - ventral subparaventricular zone
VLPO - ventrolateral preoptic nucleus
MA - medial amygdala
LHb - lateral habenula
LGv - ventral lateral geniculate nuclei

FIGURE 3.2   Targets of ipRGCs in the brain.

- in the shell of olivary pretectal nucleus (OPN), a center for the pupillary light reflex,
- in the ventral subparaventricular zone (vSPZ), implicated in acute arrest of locomotor activity by light in nocturnal animals,
- in the posterior thalamic nucleus, dorsal border, nociception,
- in the ventrolateral preoptic nucleus (VLPO), a control center for sleep,
- in the medial amygdala,
- in the lateral habenula (integration of the limbic, motor, and circadian systems).

Schmidt et al. [2011b] suggested that M1 ipRGCs projecting to the OPN are different from the M1 subtype projecting to the SCN, hinting at the possibility of subpopulations within the M1 cells.

In numerous functions, atypical melanopsin cells cooperate with the classical photoreceptors, rods and cones. The recently discovered ipRGCs integrate their intrinsic, melanopsin-mediated light information with rod/cone signals relayed via synaptic connections to influence light-dependent behaviors. However, even in the absence of signals from rods and cones (experimental knockout rodless/coneless mice), ipRGCs are able to drive a variety of behaviors, some with a surprising degree of perfection [Do et al. 2010].

Mechanisms in which melanopsin cells control light-dependent behavior were the subjects of various studies. Some data suggest that melanopsin plays an important role in providing information about spatial brightness to the thalamo-cortical visual projection and, perhaps, other aspects of vision [Schmidt et al. 2011a].

Stabio et al. in their 2018 work showed a new role and a new possible approach to M5 cells. They have confirmed their photosensitivity, but also described their role in visual function, suggesting that M5 cells are a morphologically and functionally distinct unique ipRGC type. M5 cells (typically associated with non-image-forming functions) also innervate the dorsal lateral geniculate nucleus (dLGN), influencing the visual cortex. M5 cells, which prefer UV over green stimuli throughout the retina, might be a source of excitatory geniculocortical drive for such UV-preferring cortical neurons [Stabio et al. 2018]. On the other hand, synaptically driven properties and brain projections of the M5 cell subtype implicate them in mechanisms of visual perception, especially color vision. Some mouse SCN neurons have been reported to exhibit cone-dependent blue-on/yellow-off spectral opponency [Walmsley et al. 2015; Stabio et al. 2018]. Chromatic cues provide a more reliable indication of time of day than changes in illumination alone. Retinal input to the mouse SCN probably derives from chromatically unselective M1 and M2 cells, whereas the intergeniculate nucleus projects to the SCN, providing a possible pathway by means of which M5 cells might indirectly supply chromatic information (selective information about the color) to the circadian clock [Stabio et. al 2018]. In humans, the melanopsin-expressing ganglion cells are present in the central retina and form a photoreceptive network of dendrites throughout the foveal specialization, but with only a few processes crossing the foveal pit. Their axons project to the LGN. These ganglion cells receive input from the spectrally distinct short (S), middle (M), and long (L) wavelength-sensitive cone photoreceptors [Dacey et al. 2005]. Projection from melanopsin cells to the dorsal LGN gives rise to irradiance sensitivity in the thalamocortical pathway and could play a role in visual perception [Brown et al. 2010].

Non-M1 ipRGCs also project to periaqueductal gray and to the amygdala, which suggests a contribution to the pain, fear, and anxiety circuitry [Hattar et al. 2006].

In conclusion, ipRGCs have numerous functional designations: their influence is seen in non-image- and image-forming vision, retinal development, migraine research, photoallodynia, sleep disorders, anxiety, and depression. A mutation of melanopsin may be associated with seasonal affective disorder [Roecklein et al. 2009]. Since ipRGCs are driven by intense light applied to the eye, particularly light enriched with blue, using such stimulation to provide effective treatment of this disease can be considered.

3.4 PHOTOTRANSDUCTION BY ipRGCS, PHYSIOLOGICAL ROLE OF MELANOPSIN, CLINICAL IMPLICATIONS OF MELANOPSIN PATHWAYS

Chemically, melanopsin is a photopigment with a signature which is an unusual tyrosine residue in the counterion position for the Schiffbase linkage between the chromophore and the opsin protein [Provencio et al. 2000; Provencio et al. 2002]. In mammals, melanopsin expression is typical for ipRGCs. Melanopsin has peak absorbance at about 480 nm (blue light). It is isomerized on light absorption, converting 11-cis retinal to all-trans-retinal such as rod and cone photopigment. Like all photopigments, melanopsin is a prototypical G-protein-coupled receptor. TPR channels are nonselective cation channels, mediated by iPRGCs phototransduction. Cyclic nucleotide-gated ion channels (CNG channels) mediating phototransduction in rods and cones are not present in ipRGCs. Melanopsin physiological response is characterized by slow temporal kinetics and sustained signaling after light cessation.

According to the anatomic projection of ipRGCs, several non-visual effects of light mediated by melanopsin are described [Spitschan 2019a; Spitschan 2019b]:

- Pupil size regulation – strongly controlled by melanopsin dynamic pupil responses to light. All photoreceptors are involved in the control of pupil size [Spitchan 2019b].
- Melatonin suppression – a hormone produced by the pineal body during the night ("night hormone"). Melatonin production is suppressed by light via the retinohypothalamic pathway connecting ipRGCs to the suprachiasmatic nucleus. In some blind people, light suppresses melatonin secretion by ocular exposure to bright white light.
- Circadian phase shifting – night exposure to light – can shift the circadian rhythms of different biological processes, which are synchronized to the external light–dark cycle via the retinohypothalamic pathway. Illumination of the retina evokes excitatory postsynaptic potentials in a subpopulation of SCN neurons. These potentials are the result of glutamate release from the retinothalamic pathway and are mediated by inotropic and metabotropic glutamate receptors. Additionally, SCN-projecting ipRGCs synthetize pituitary adenylate cyclase-activating polypeptide (PACAP), which may act as a modulator of the glutamatergic input to the SCN [Sollars et al. 2015; Hannibal 2006]. Circadian phase-shifting is most spectacular for short wavelength light fitting melanopsin activation. In humans, cones seem to contribute to phase-shifting, and this contribution depends on the timing of the light exposure [Gooley et al. 2010; Spitschan 2019a]. It was also shown that rods may contribute to melatonin suppression [Vartanian et al. 2015].

Some data suggest that melanopsin contributes to visual perception. In primates, melanopsin-expressing ganglion cells signal color and irradiance and project to the lateral geniculate nucleus of the thalamus (a gate controlling the passage of information to the primary visual cortex in the occipital lobe) [Dacey et al. 2005]. In humans, the mechanism is different: pulses of light that only stimulate melanopsin elicit activity in the primary visual cortex [Spitschan et al. 2019a]. Other data emphasize the role of melanopsin signals in the detection and discrimination of light, temporal and color processing [Zele et al. 2018a], brightness estimation [Zele et al. 2018b], and color perception [Cao et al. 2018]. In mouse vision, melanopsin-driven light adaptation was described [Allen et al. 2014]. Spitschan et al. [2017] as well as Allen et al. [2019] demonstrated that melanopsin may contribute to spatial vision as a "raumgeber" ("space-giver").

These data provide grounds for scientists to consider all retina cells and their functional interaction as a place of complex integration of vision science and chronobiology (chronomedicine) as well as a place of interaction of image and non-image vision.

The retina is not the only tissue that contains melanopsin; this photopigment has been found in the iris, the ciliary body, and in blood vessels [Marek et al. 2019].

Previously, melanopsin-expressing ipRGCs were also thought to mediate light-induced pain. Some data showed the possibility of an alternative light-responsive pathway independent of the retina and optic nerve. Matynia et al. discovered that in humans and mice, melanopsin is expressed in 3% of small trigeminal sensory ganglia neurons (TG), localized in the ophthalmic branch of the trigeminal nerve (containing nociceptors), and reported their intrinsic photosensitivity [Matynia et al. 2016]. As described above, ipRGCs respond to light in ways that are intrinsic (from melanopsin) and extrinsic (from rod and cone photoreceptors via their inner retinal circuits). TG neurons do not receive sensory inputs from other neurons, but respond to blue light stimuli with a delayed onset and sustained firing, similar to ipRGCs [Matynia et al. 2016]. The identification of an active photopigment, melanopsin, in the classic trigeminal circuitry provides evidence for a novel interpretation of specific pathophysiological conditions, i.e. photoallodynia, migraine-related photophobia, and for interaction between visual and trigeminal sensory processes. In humans during photophobia periods, Moulton et al. [2009] reported fMRI-detected activation at the level of TG, the trigeminal nucleus caudalis, and the ventroposteromedial thalamus. In 2019, Marek et al. demonstrated that blue-light exposure provokes important immune and inflammatory responses in the ocular surface, in trigeminal pathways and the retina, and/or aggravates dry eye disease. What was considered was the possibility that mechanisms involving melanopsin transmit phototoxic messages from the retina to the trigeminal system:

- the ipRGCs-independent alternative pathway of light avoidance [Matynia et al. 2016],
- intra-retinal processes, independent of central visual centers, producing an enhanced trigeminal response to light [Dolgonos et al. 2011],
- probably the roles of the iris and the ciliary body: the iris would be able to receive and transmit the photic signal in the trigeminal system [Marek et al. 2019] and melanopsin is expressed in the ciliary body itself [Semo et al. 2014].

Further studies of retinal and near-retinal interactions with the trigeminal system would make it possible to better understand blue-light aversive behavior mechanisms and develop better treatment for photophobic patients.

In addition to some clinical implications of the melanopsin-based non-image-forming visual system described above, molecular studies and animal models emphasized various new possible roles for the non-image-forming visual system in associative learning, metabolic control, and image formation [Ksendzowsky et al. 2017].

3.5 THE NEUROBIOLOGY OF CIRCADIAN RHYTHMS – DO WE NEED LIGHT?

The paired suprachiasmatic nuclei (SCN) located in the anterior hypothalamus, dorsal to the optic chiasm, are the primary circadian oscillator in the brain. Each nucleus is composed of about 10,000 interconnected small neurons, expressing circadian rhythmicity in the rate of action potential firing and in the gene expression. Several animal studies confirmed that SCN is the site of a circadian oscillator whose neurons

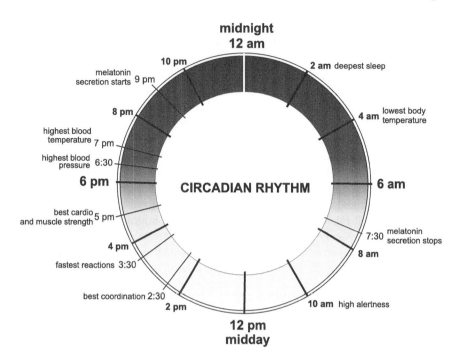

FIGURE 3.3   Circadian rhythm of human biological activity.

rhythmically alter their metabolism and activity/firing rate [Sollars et al. 2015]. Retinal projection terminating in the SCN (the retinohypothalamic tract) enables direct synchronization with the light/darkness cycle and entrainment to the environmental day/night cycle (entrainment means more than synchronization – it allows for great plasticity and adaptation) (Figure 3.3). The ipRGCs – melanopsin photoreceptors located at the beginning of the retinohypothalamic tract – have spectral characteristics: blue light (480 nm) is the strongest stimulus, whereas red light (≥600 nm) has a minimal effect on melanopsin response. Daytime sunlight contains more blue wavelengths than sunset. As the sun reaches the horizon, short wavelengths are scattered in the atmosphere and longer, redder wavelengths more easily reach the surface of the earth [Bedrosian et al. 2017]. These authors suggest that "the sensitivity spectrum of melanopsin may be an adaptation to the natural solar cycle, so that ipRGCs are tuned to discriminating daylight from evening, better entraining the circadian rhythm". Light detected by ipRGCs sets the molecular clock in the SCN.

The molecular substrate of SCN cells responsible for their autonomous circadian oscillations, which are generated by two transcription/translation feedback loops producing circadian rhythms of clock gene expression, are four integral clock proteins and enzyme kinases and phosphatases regulating the localization and stability of clock proteins. There are:

- two activators: CLOCK and BMAL1 (circadian locomotor output cycles kaput and brain and muscle ARNT-like 1, respectively),
- two repressors: PER and CRY (period and cryptochrome, respectively).

It seems that the intracellular molecular clock drives the expression of rhythms in the frequency of action potential firing in SCN neurons [Brancaccio et al. 2014; Sollars et al. 2015]. To understand the physiological role of light in human activity, it is particularly important to note the following observations:

- light exposure during the subjective day (normal exposure to daylight) has no effect on the phase of the circadian rhythm (free-running circadian rhythm),
- light exposure early in the subjective night delays circadian rhythm (delay phase),
- light exposure late in the subjective night advances circadian rhythm (advance phase).

Activation of the retinohypothalamic tract stimulates SCN neurons, and increases their spike rate and intracellular calcium levels. The light-evoked increase in the intracellular calcium level seems to play a role in shifting molecular oscillations [Diekman et al. 2013]. The resistance of the clock gene induction (*per1, per2*) and the clock to the phase-shift during the subjective day (the low effect of light on the increase of the neuron spike rate in the SCN) may be due to the fact that intracellular calcium levels are stabilized at a high level during the day.

Abnormalities in the clock gene function may be a cause of mood disorder pathology.

SCN neurons receive a serotonergic input from ascending projections of serotonergic neurons in the median raphe nucleus. Serotonin inhibits the response of SCN to light via presynaptic serotonin 1B receptors on retinothalamic nerve endings, decreasing the ipRGCs' glutamate release [Sollars et al. 2015]. Serotonin activity and the function of its 1B receptors seem to be involved in the pathophysiology of seasonal affective disorder, winter depression, or depression-like states.

SCN controls rhythms in behavior, physiology, and hormone secretion. Anatomical pathways exiting SCN are responsible for the following functions:

- consolidation of wakefulness (influence on the hypocretin/orexin system),
- regulation of many autonomic functions, including rhythmic pineal melatonin secretion (melatonin – the "darkness hormone", a small indoleamine that is produced and secreted in a 24 h rhythm that peaks at night; its secretion is suppressed by light),
- contribution to the circadian rhythm of corticosterone and cortisol secretion,
- projection to the ventrolateral preoptic nucleus, promoting sleep,
- projection to the subparaventricular zone – relay for SCN signals, involved in the control of multiple circadian rhythms, including rhythmic sleep, body temperature, and locomotor activity.

SCN is not the only circadian oscillator. Several regions of the brain, all organs of the body, and most cells throughout the body contain autonomous circadian oscillators, using the same clock genes that are active in the SCN. Peripheral clocks require external and internal signals, including signals from the central clock, to synchronize their circadian rhythms [Zhao et al. 2019]. The SCN-regulated daily rhythm in glu-cocorticoid secretion and rhythmic regulation of glucocorticoid receptors seem to be a signal maintaining synchrony among central and peripheral oscillators and gluco-corticoids, which seem to be potent transcriptional regulators [Sollars et al. 2015]. The circadian system regulates glucocorticoid secretion from the adrenal glands, so that the concentrations tend to peak in the morning just before awakening and decrease throughout the day. Glucocorticoid dysregulation is associated with a num-ber of mood disorders, and hypercortisolemia is detected in major depression patients [Wolkowitz et al. 2009; Bedrosian et al. 2017]. Light directly affects the secretion of glucocorticoids in humans. Circulating melatonin is involved in the entrainment of clocks located in peripheral organs through interactions with the molecular clock mechanism, i.e. the phase-resetting clock genes [Bedrosian et al. 2017]. Thus, light exposure interacts with the hypothalamic–pituitary–adrenal axis. During the night and in darkness, melatonin concentration rises and promotes the onset of sleep. Exposure to light during the night strongly suppresses melatonin, worsening sleep quality. Indoor light pollution, but also light pollution outside the home (areas with more outdoor night lighting) are sufficient to disrupt sleep (less quality and quantity of sleep, daytime sleepiness), contributing to depressed mood [Ohayon et al. 2016; Koo et al. 2013; Palmer et al. 2017]. Stephenson et al. [2012] directly suggested that light, via melanopsin, may exert its antidepressant effect through a modulation of the homeostatic process of sleep, working as a photic regulation of mood.

In summary, the following conclusions should be formulated:

- temporal organization and synchronization of biological processes are criti-cal for human health,
- aberrant light exposure can depress mood, metabolism, and the immune system, and increase cancer risk, perturbing synchronization of the central biological clock in the SCN with peripheral clocks throughout the brain and body [Bedrosian et al. 2017].

3.6 THE ALERTING EFFECTS OF LIGHT

In terms of cognitive-behavioral performance, "alertness is also used to describe a state of vigilance or sustained attention in which the person is able to achieve and maintain a certain level of cognitive performance while executing a given task" [Łaszewska et al. 2018].

There are subjective (high subjective alertness ratings, low subjective fatigue rat-ings), and objective (behavior and brain activity) correlates of alertness [Łaszewska et al. 2018]. In addition to the results of performance tests, reliable neurophysiologi-cal measurements of human alertness are used, i.e. electroencephalographic (EEG) measurements, with low power densities in the theta [Cajochen et al. 2005] and alpha frequency range [Figueiro et al. 2009]; electro-oculographic (EOG) slow rolling eye movements; and eye blink rate [Cajochen et al 2005]. Alertness is associated with levels of melatonin and core body temperature [Figueiro et al. 2009]. Its level may be considered in the context of the classic two-process model of sleep regulation: it is influenced by circadian rhythm and homeostatic sleep pressure. The additional process encompasses endogenous (chronotype) and exogenous (physical activity, food intake) factors and other afferent processes integrating sensory inputs from body system factors, which are important for sleep regulation. Light plays an essential role in maintaining the level of alertness. The alerting effects of light depend on its spectral wavelength, duration, and intensity [Xu et al. 2018]. The systematic review study by these authors aimed to document factors influencing the alerting effect of light, and study effective and non-effective light intervention. They showed that the minimum light intensity required to induce an alerting effect is higher during the day than at night, and this minimum light intensity varies with the spectral distribution of light. They also concluded that blue light of low irradiance is an effective light intervention for increasing alertness levels at night, but is less effective during the daytime. Moderate bright white light seems to be effective in reducing sleepiness during the daytime, but is less effective at night. They also emphasized the role of environmental factors (like prior light exposure) and individual factors (chronotype, several activities undertaken during and just before the measurement of sleepiness) in the evoked alerting effect of a light intervention. Our previous study showed that long-wavelength light induced a strong response of bioelectrical activity during the daytime, which is more spread over the scalp, suggesting the role of visual mechanisms in affecting daily alertness [Łaszewska et al. 2017]. These data might be important for the development and use of proper light infrastructure and light interventions for different home and workplace settings.

3.7 SUMMARY

The discovery of the third class of retinal photoreceptor ganglion cells – intrinsically photosensitive retinal ganglion cells (ipRGCs) – launched research into non-image vision. This small population of cells is an atypical photoreceptor group. They express a new photopigment: melanopsin. In mammals and in humans, ipRGCs drive a variety of non-image visual signals and modulate a multiplicity of physiological functions and behaviors: circadian photoentrainment, pupillary light reflex, activity masking, sleep/arousal, neuroendocrine activity, anxiety, and light aversion. In numerous functions atypical melanopsin cells cooperate with the classical photoreceptors, rods and cones. The diversity of ipRGC subtypes (M1–M5) emphasizes the complex organization of the non-visual system on the photoreceptor level. The ipRGCs located at the beginning of the retinohypothalamic tract have spectral characteristics; blue light (480 nm) is the strongest stimulus, whereas red light ($\geq$600 nm) has a minimal effect on melanopsin response. The sensitivity spectrum of melanopsin may be an adaptation to the natural solar cycle, so that ipRGCs are tuned to discriminating daylight from evening, thus better entraining the circadian rhythm. Light detected by ipRGCs sets the molecular clock in the SCN.

Understanding the physiological mechanisms of the non-visual system may be important for the development and use of proper light infrastructure and light interventions for different home and workplace settings, especially for shift work conditions.

4 Lighting Quality and Well-Being

> There are two kinds of light – the glow that illumines, and the glare that obscures.

James Thurber

4.1 WHAT DO WE MEAN BY LIGHTING QUALITY AND WELL-BEING?

4.1.1 WELL-BEING

There are many definitions of well-being. Merriam-Webster's Dictionary [Webster 2019] defines it as the experience of health, happiness, and prosperity. This includes having good mental health, high life satisfaction, a sense of meaning, and the ability to manage stress. More generally, well-being is just feeling well. The World Health Organization defines health and well-being together: "Health is a state of complete physical, mental and social well-being and not merely the absence of disease or infirmity" [WHO 2020]. Attention is drawn to the fact that it cannot be just a lack of diseases or other negative feelings. For example, the absence of disease is not enough to talk about well-being. In addition, this issue is considered taking into account relations with the social and work environment in a cultural and demographic context. In this way, five major aspects of well-being are discussed: emotional, physical, social, workplace, and societal. On the other hand, it is an individual and subjective state. Under some given conditions, not everyone feels the same.

Light has accompanied humans since the dawn of time. The need for light is obvious to everyone. However, not everyone realizes how strong its impact is on well-being. Workers who evaluate lighting well will also assess the room as more attractive. This means that they will start work in a better mood, achieve greater satisfaction with their tasks, and go home in a good mood. However, for this state of affairs to occur, good quality lighting is needed.

4.1.2 LIGHTING QUALITY

The times when the main purpose of illuminating the workplace was to provide an appropriate luminous environment in which, through the sense of vision, people can function effectively, efficiently, and comfortably [Rea 2000] are gone forever. Today we know that lighting affects people and their health in both visual and non-visual ways, which was described in Chapters 2 and 3. Besides the visual effects, the emotional and biological impact of light on humans is also considered. Along with the development of knowledge about the influence lighting has on people, the approach to the design and assessment of lighting has changed. This new, multifaceted perspective is frequently referred to as human-centric lighting (HCL – described in Chapters 7 and 9). Nevertheless, the visual effect has remained very important, and development is taking place in this area as well. All the lighting factors and parameters that affect the visual effects of light that are understood as ensuring visual performance and comfort can be classified as determining the quality of lighting, i.e. we are talking about lighting quality.

Light sources, luminaires, and methods used to design, control, and evaluate lighting in workplaces are now sophisticated, and interesting solutions are being introduced that require good knowledge of lighting engineering. For this reason, the problem of assessing lighting quality is very complex. There are many different approaches to such estimation and many attempts to parameterize it. In recent years, the important role of energy consumption has also been emphasized, although this issue is more related to the economics and technology of lighting. Although we appreciate the importance of this problem, it will not be described in our book.

Lighting quality in the context of visual comfort is often considered by negation [Boyce 2014]: good quality of lighting favors visual comfort when there are no discomfort factors. Indeed, the description of discomfort could be easier. However, it seems that it is worth trying to specify our expectations from high quality lighting.

In this chapter, we will present only those lighting quality parameters that are primarily associated with the visual effects of light, and with those emotional/psychological (e.g. glare, color appearance) or biological (e.g. flicker, stroboscopic effect) effects that result from the appropriate/inappropriate use of those parameters of lighting quality.

Lighting quality in indoor workplaces is considered, taking into account several independent aspects. It is assumed that lighting should:

- ensure safety: it should provide conditions for people to safely move, perform activities and recognize dangerous situations,
- ensure the visual discrimination of details: shape, size, and contrast around the task being performed and, in this way, make it possible for work to be performed effectively and accurately,
- assist in creating a luminous environment, which promotes visual comfort and well-being. Workers spend many hours in the same workplace every day. Lighting is one of the important factors determining their well-being resulting in their job satisfaction, motivation, and commitment to work, as well as promoting work efficiency and decreasing absenteeism or employee turnover,
- favor alertness and not disturb the circadian rhythm through non-visual effects (these aspects are presented in Chapters 7 and 8).

Usually several areas that relate to the quality of lighting are considered. These are: lighting intensity, spatial distribution of light, color aspects, flicker, and stroboscopic effect (see Table 4.1). They will be presented in this order in the subsections that follow.

TABLE 4.1

The Area of Interest and Related Lighting Parameters Actually Covered by Lighting Standards (Lighting Quality Related to Visual Effect of Light)

Area of interest	Parameter	Measurement/assessment area
Lighting intensity	Average illuminance	Work plane, task area, walls and ceiling
Spatial distribution of light	Illuminance uniformity	
	Discomfort glare – unified glare rating	Luminaires in the visual field of workers
	Reflectance	Main interior surfaces (walls, ceiling, floor, furnishings etc.)
	Cylindrical illuminance (average vertical plane illuminance)	Activity area, vertical/horizontal plane, at the height of occupant's
	Modeling – ratio of cylindrical illuminance to horizontal illuminance at the point	eyes (1.2 m or 1.6 m above the floor)
Color aspects	Color appearance (correlated color temperature)	Light sources installed in luminaires
	Color rendering	
Flicker	–	Lighting installation/luminaires
Stroboscopic effect	–	

Based on EN 12464-1 Standard [2011].

4.1.3 STANDARDIZATION CONCERNING LIGHTING DESIGN

4.1.3.1 International/European Lighting Standards

The lighting standards which are currently in force which were published by such world recognized and honored organizations as the European Committee for Standardization (CEN), the International Standard Organization (ISO), and the International Committee on Illumination (CIE). They focus on providing lighting related only to the visual effects of light (Table 4.1). The standards include lighting requirements that aim to provide good quality lighting and ensure "visual comfort and performance of people having normal ophthalmic capacity" [EN 12464-1]. The main lighting standard with requirements and recommendations for lighting design used in Europe is EN 12464-1 [2011]. Some associated documents outside the EU are, for example, ISO 8995-1 [2002] and CIE S 008/E [CIE 2001]. As stated in the scope of EN 12464-1, this standard "does not specify requirements with respect to the safety and health of people at work". However, there are two very short subchapters in this standard, which indicate the additional benefits of daylight lighting and mention that the variability of light is important for human health and well-being. Still, no specific indications or guidelines are given. In general, it can be said that the existing lighting standards do not provide any meaningful guidelines for designers, even in the era of the growing necessity to consider health, well-being, and performance in lighting design.

It is worth mentioning that the international standardization organizations (ISO, CEN, CIE) are working on new standards which should develop a method for assessing and planning lighting while taking into account the visual and non-visual effects of light:

- the Technical Committee of the International Organization for Standardization ISO/TC 274 Light and Lighting has developed a new standard: ISO/WD TR 21783 Light and lighting – Integrative lighting – Non-visual effects,
- the Joint Technical Committee of CIE and ISO (CIE-ISO): JTC 14 Integrative lighting has developed requirements for integrative lighting,
- the Technical Committee of the European Committee for Standardization CEN TC169 Light and Lighting is working on updating the standard EN 12464-1 [2011].

It is worth mentioning that both Technical Committees: ISO TC274 Light and Lighting and CEN TC169 Light and Lighting develop application standards in lighting based on CIE fundamentals in the field. We should, therefore, expect that the new standards developed by these committees will be consistent with each other in the new lighting requirements. However, it would be good if, parallel to the new lighting requirements, measurement verification procedures and computer programs for interior lighting design (or supplementing actual lighting design software, like Dialux, Relux etc.) could be developed. Until then it would be advisable to try to deal with the existing recommendations for human-centric lighting in a different way. Taking into account the existing recommendations focused on the visual effect of light, it seems reasonable to:

- ensure compliance with the standard requirements for qualitative and quantitative lighting parameters,
- strive to introduce basic recommendations for integrative lighting (human-centric lighting) design.

4.1.3.2   Standard for Building Design

WELL Building Standard is a multi-criteria building certification system created by the International WELL Building Institute™. This is a system which promotes the health and well-being of building users. Its certification gives confirmation to building owners and employees that the way they design their space promotes health and well-being. It consists of a number of performance indicators, and design and implementation strategies that regulate various aspects affecting the health, comfort, and awareness of users. These aspects are grouped into ten categories: Air, Water, Nutrition, Light, Movement, Thermal Comfort, Acoustics, Materials, Mind, and Community.

The category *Light* includes guidelines for both visual and circadian lighting design. The Institute's standard presents guidelines for artificial as well as natural lighting in interiors. The latest issue of the 2019 WELL Standard [WELL 2019a] contains recommendations for four indoor categories: work areas, living environments, break rooms, and learning areas.

The logical arrangement of lighting recommendations deviates from the typical layout in existing lighting standards, especially in the area of visual lighting design. Some of the lighting parameters commonly attributed to visual lighting design, like glare control, color aspects of light sources, and reflectance, are not included in the visual lighting design area. They are discussed separately in this standard (see Table 4.2).

This standard adopted equivalent melanopic lux (*EML*) as a measure of non-visual effect of light on humans, which is based on the ipRGC-influenced action spectrum proposed by Lucas [Lucas et al. 2014]. It also describes the method of calculating the *EML*.

Providing adequate access to daylight in the interiors that are being designed is an important area of the WELL standard, which is in line with human-centric lighting (HCL) assumptions.

Recommendations for circadian lighting design are presented in Chapters 7, 8, and 9.

4.2 LIGHTING INTENSITY

The only lighting parameter determining lighting intensity is illuminance (*E*), or rather its average value in a given area. According to the vocabulary definition of lighting [CIE 2016A], illuminance (at a point of a surface) is the quotient of the luminous flux incident on an element of the surface containing the point by the area of that element.

Allowing people to perceive the features of a human face was assumed to be the minimum criterion for the appropriate level of illuminance. In interiors not intended for work, it is most important to get the overall impression created by lighting. In order to be able to perceive the features of a human face, room luminance should be around 1 cd/m^2. This can be obtained under normal lighting conditions, with an illuminance level of around 20 lx, which is therefore considered to be the minimum value for non-operational rooms.

Perception of the features of a human face occurs without much effort with luminance in the range of 10–20 cd/m^2, provided that the surrounding background is not black. This means that the illuminance on the vertical plane must be at least 100 lx and on the horizontal plane about twice as much. Therefore, horizontal illumination of 200 lx is considered a minimum that can be accepted in rooms where people stay for a long time (regardless of what visual task is being performed).

The level of illuminance needed to perform specific visual work is selected depending on the degree of difficulty of the visual task and the apparent size of the details involved. This value is determined by the reflectance of the task and the contrast of the detail of the object with its background. The lower the reflectance and the contrast, the greater the degree of difficulty of visual work.

Standard EN 12464-1 [2011] specifies the requirements for the maintained illuminance (*E$_m$*) on the reference surface for the interior (area) and the task or activity, i.e. the illuminance value on the working plane for a specific type of work. Providing the correct parameters in this sphere enables comfortable recognition of the details of a visual task. The standard [EN 12464-1] defines illuminance requirements for

TABLE 4.2

WELL Building Standard Guideline Areas Related to Natural and Artificial Lighting (Chosen for Working Areas Only)

Light feature	Focus	Parameter
Visual effect lighting design	Visual acuity	• Average illuminance on horizontal plane

	Brightness management	• Brightness contrast
		   • in the room: between task surfaces and adjacent surfaces, task surfaces and remote surfaces
		   • between main rooms and ancillary spaces
		   • across the ceiling

Circadian lighting design	Melanopic light for work areas	• Equivalent melanopic lux (*EML*) on vertical plane 1.2 m above the floor (to simulate the view of the occupant)
Electric light glare control	Luminaire shielding	• Shielding angles of luminaires in relation to light source luminance ranges
	Glare minimization	• Luminance of luminaire at more than 53° above the center of view (at workstation) • *UGR* at workstations and sitting areas
Solar glare control	View window shading	–
	Daylight management	–
Low glare workstation design	Glare avoidance	–
Color quality	Color rendering index	Color rendering index: R_a and R_9
Surface design	Surfaces reflectivity (vertical surfaces, ceiling, and furniture)	Light reflectance values – LRV (reflectance)
Automated shading and dimming controls	Automated sunlight control	–
	Responsive light control	–
Right to light	Lease depth	Workstation distance from windows
	Window access	
Daylight modeling	Healthy sunlight exposure	• Spatial daylight autonomy* • Annual sunlight exposure**
Daylight fenestration	Window size	• Window–wall ratio • % window area at min 2.1 m above the floor
	Window transmittance	Visible transmittance (VT)
	Uniform color transmittance	VT of wavelengths between 400 and 650 nm

* sDA – Spatial daylight autonomy – a new daylight metric introduced by LEED v4 – it defines a % of the area that meets minimum daylight illuminance levels for a specified fraction of the working hours per year.

** ASE Annual sunlight exposure – a new daylight metric introduced by LEED v4 – it identifies the potential for visual discomfort in interior work spaces introduced by LEED v4; no more than 10% of a space should have direct sunlight of more than 1000 lx for a maximum period of 250 hours per year.

the task area, the immediate surrounding area (a band of at least 0.5 m surrounding the task area within the visual field), and the background area (at least 3 m width adjacent to the immediate surrounding area within the limits of the space).

However, for the choice of adequate illuminance level, two independent factors are often highlighted: individual preferences and the luminous environment's impact on working conditions. The first large-scale experiments concerning the relation between illuminance level and observers' preferences were carried out by seven researchers in the years 1957–1968 [de Boer and Fischer 1978]. In total, nine experiments were conducted in the office environment on a group of 1,930 participants. In the experiments, the illuminance level of the room was changed. After completing a visual task, each participant assessed the quality of lighting on a horizontal work plane (on the scale: too dark – satisfactory – too bright). The results of the analysis of these studies were first described in 1978 [de Boer and Fischer 1978]. A graph showing the relationship between the number of individual ratings and the illuminance level is shown in Figure 4.1. The very important conclusion of those studies was that individual preferences are very difficult to satisfy – there is no level of illuminance at which everyone would be satisfied. The largest group of participants preferred illuminance at the level of approximately 2000 lx – it is worth noting that this value is much higher than 500 lx – the illuminance level recommended in the standard for office work. Later studies have shown that other factors also have a very strong impact on users' satisfaction with lighting. First of all, the type of work performed [van Ooyen et al. 1987] and the illumination of the surroundings of the task area (or room lighting in general). In another paper [Manov 2007], attention was paid to the very important role of contrast between the task area and the background. The impact of the environment (the luminance uniformity of walls) is analyzed by Chraibi et al.

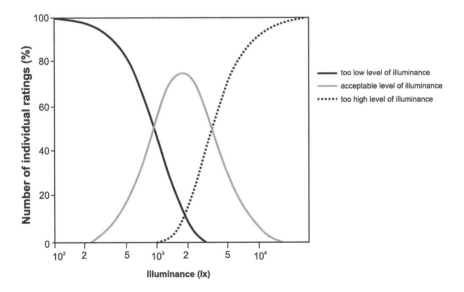

FIGURE 4.1   Preferred illuminance in the working environment – a number of individual ratings (in %) as a function of illuminance level (E).

[2017]. The authors analyzed the impact that two different luminance distributions on the wall – but of the same average luminance (about 200 cd/m[2)] – had on the workers' preference for illuminating the task area on their desk. On one wall, light was distributed non-uniformly (min.: 48 cd/m^2, max.: 600 cd/m^2), while on the other more uniformly (min.: 94 cd/m^2, max.: 362 cd/m^2). In the first case (non-uniform luminance distribution), participants preferred a lower level of desk illuminance than if luminance distribution was uniform. On the other hand, it has been shown that for an average wall luminance of 50 cd/m^2 or 100 cd/m^2, uniform luminance distribution on the wall does not affect the user's choice of illuminance level of desk lighting. The authors of the same work [Chraibi et al. 2017] suggest the possibility of reconciling differences in the preferences resulting from individual predilections. The choice of illuminance level could be affected by different illuminance levels when beginning the experiment (the so-called start level). Lower start level illuminance (300 lx) led to participants' later choice of an illuminance level as one that was optimal for them that was lower than if higher illuminances were used for the start level (500 lx or 700 lx). As a result, smaller individual differences in workplace lighting preferences were achieved. Current research published in 2019 [de Bakker 2019] confirms the problem of individual preferences (also in the context of dynamic lighting). The authors suggest that if possible, illuminance level (and changes in dynamic lighting) should be selected individually, according to the preferences of particular employees.

The influence of the type of task on illuminance preferences has been shown using the example of a document task performed only on paper versus a computer task [Lee et al. 2014]. The authors examined the acceptance of illuminance for various lighting conditions in office conditions. The results showed that the change in the type of task from a paper document to the computer was associated with less demand for illuminance level in the horizontal plane for work on the computer.

The choice of illuminance level is strictly related to the biological effects of light. It was proven that 500 lx at the horizontal plane during the biological day means biological darkness for humans and could result in circadian rhythm disruption. These aspects of illuminance level are considered in Chapters 7 and 8.

The illuminance level for artificial light on the working plane in indoor workplaces is determined by national and international standards or recommendations. It usually ranges from 200 to 750 lx (but the illuminance range in the EN 12464-1: 2011 standard is between 20 lx and 5000 lx, and the illuminance level is selected from the illuminance scale with a gradation factor of 1.5). Moreover, it also varies depending on the type of work, e.g. 300 lx may be recommended for copying or reception desk work, but 500 lx for writing, typing, reading, data processing. However, these are the so-called minimum requirements to be met in order to provide visual comfort for efficient work performance, while meeting the aspects of energy efficiency. In comparison with the intensity of daylight illumination, these are very low values and for the non-visual effects of light, they constitute biological darkness.

Independently conducted studies have shown that in the case of most jobs, it is worth raising the level of illuminance in the task area. Increasing the illuminance to 1000–2500 lx causes a visible increase in task performance (by 7%–30%) and reduces the number of rejections (by 18%–29%) [van Bommel et al. 2002]. A slight increase in horizontal illuminance from 450–600 lx to 800–1300 lx and vertical illuminance from

100–300 lx to 250–500 lx resulted in an increase of performance at various workstations by up to 7% [Juslen 2007]. It is worth adding that the recommended level for proper synchronization of circadian rhythms is 2000 lx [Preto et al. 2019]. Therefore, why not increase the value in the standards similarly to the illuminance at indoor workplaces? The only justification is the economic factor – increasing the illuminance is associated with higher electricity costs and we want to save energy. Unfortunately, in this case, these are sometimes savings at the expense of work efficiency.

4.3 SPATIAL DISTRIBUTION OF LIGHT

Spatial distribution of light is related to the following lighting parameters: uniformity of illuminance, glare, the reflectance of main surfaces, modeling, cylindrical illuminance. These will be discussed below.

4.3.1 LUMINANCE DISTRIBUTION

Luminance distribution in the visual field is responsible for the eyes' adaptation level and affects the visibility of the task. It is important to create balanced luminance distribution which positively affects visual acuity, contrast sensitivity, and such visual functions as accommodation, convergence, eye movements etc. Besides, it promotes visual comfort.

Luminance distribution largely determines the mood in the interior and its decorativeness. A surface with a high luminance value seems to be more distant than a surface with a low luminance value. Thus, bright walls give the impression that the room is larger, and dark walls visually reduce the space, just as a light ceiling appears to be higher than a dark one.

Humans need luminance contrast so that they can perceive the surrounding environment. The greater the contrast, the more easily perceived the object, e.g. black letters on a white background. In cases when contrast is low, lighting intensity must be increased to guarantee easy identification of the object. One should be aware of the fact that the visual result will not be satisfactory in every situation.

Luminance distribution in the interior is determined by specifying the luminance ratio characteristic for the chosen fields. To this end, fields appearing in the user's close and distant environment are distinguished. The luminance ratio CR is calculated according to Formula (4.1) as the ratio of the luminance of the object (L_o) to the luminance of the background (L_b). Average luminance ratios are calculated for the object and planes (adjacent and not adjacent) (e.g. document – table top) and the planes to which eyesight is moved (e.g. from the document to the computer keyboard).

$$CR = \frac{L_o}{L_b} \tag{4.1}$$

Usually the numerical notation of the CR is written as a fraction. Instead of writing a fraction as 1/X, the 1:X form is used, which is more pictorial and useful in practice. This term was used in various standards and lighting recommendations to assess the luminance distribution in the field of view. The best-known principle for achieving a not-glaring, balanced distribution of luminance in a stationary field of view (e.g. at an office workstation) is not to exceed the following values: 1:3:10 (visual task: adjacent surfaces: non-adjacent surfaces). In conditions of office work, in using a computer, and in many industrial rooms, it is not possible to meet the recommended ratio of 1:3. Therefore, based on the results of our own research, it can be assumed [Wolska et al. 1999] that luminance contrasts may exceed the value of 1:3 without significantly changing visual fatigue, but the maximum allowable luminance ratio should be kept below 1:10 for the visual task-adjacent surface.

For a given level of illuminance, differences in luminance result from differences in the reflection of the light beam from the surfaces. Although illuminance may be appropriate for a particular visual task, it does not necessarily have to give a satisfactory distribution of luminance in the interior as a whole.

The other way to indirectly provide balanced luminance distribution is to ensure that the recommended reflectance values of the individual surfaces in the room (assuming diffuse reflections) are providing the right illuminance level for the walls and ceiling. In rooms intended for work, it is recommended that the average reflectance should be as follows [EN 12464-1]:

- ceiling – from 0.7 to 0.9,
- walls – from 0.5 to 0.8,
- work surfaces – from 0.2 to 0.7,
- floors – from 0.2 to 0.4.

Once the reflectances have the recommended ranges, the next step is to provide the minimum illuminance on the wall: at least 50 lx (and illuminance uniformity of at least 0.1) and on the ceiling at least 30 lx (and illuminance uniformity of at least 0.1) [EN 12464-1].

4.3.2 Discomfort Glare at Indoor Workplaces

Glare is a phenomenon accompanying the process of vision which may cause a feeling of discomfort or reduction in the ability of human perception. Glare is a certain state or vision process in which there is a feeling of discomfort or a decrease in the ability to recognize objects due to improper distribution of luminance or excessive contrasts in space or time. This can be caused by bright light sources appearing in the human field of view (direct glare) or by reflections of glare sources from directional reflecting surfaces (reflected glare). Too large a range of luminance values can cause two types of effects: a feeling of discomfort (discomfort glare) or limitation of perception (disability glare).

According to the international lighting dictionary published by the CIE in 1987 (and the revised version of 2011 [CIE 2016A]), two main types of glare are distinguished: discomfort and disability, with the following definitions.

- discomfort glare – causes discomfort, without necessarily impairing the vision of objects.
- disability glare – impairs the vision of objects without necessarily causing discomfort.

Glare is a phenomenon that is very difficult to assess. The analysis of glare and its impact on human psychophysiology is the starting point for understanding the different types of glare and the assessment methods associated with them. To facilitate the process of this phenomenon, a measure has been proposed – the glare index. This measure reflects the most important glare properties. Unfortunately, the complexity of the phenomenon and the difficulties in its description resulted in the creation of many different indexes assigned to specific conditions (indoor workplaces and artificial lighting, outdoor workplaces and artificial lighting, road lighting, indoor workplaces and daylight). It is worth paying attention to the numerous groups of glare indexes – many of them relate to the same or similar lighting conditions. Practically all of them are still being developed in various research centers. On the one hand, the diversity of the indexes and the lack of one universal index demonstrate the need for extensive unification. On the other hand, however, it hinders their practical application for evaluation and enforcement of correctly implemented lighting installations. However, this situation provides an opportunity to conduct very interesting research.

It is assumed that if there is limited discomfort glare in interiors, disability glare is also limited. For this reason, the assessment of glare in the interior is usually reduced to the assessment of discomfort glare. Unified Glare Rating (UGR) is the discomfort glare index for indoor workplaces. It was defined in CIE documents on the basis of experimental studies [CIE 1983; CIE 1995b]. The rules of application are described in the standard [EN 12464-1]. The method of calculating the UGR value is defined in Formula (4.2)

$$UGR = 8\log_{10}\left(\frac{0.25}{L_u} \sum_i \frac{L_i^2 \omega_i}{P_i^2} \right) \tag{4.2}$$

where:
L_u – background luminance (cd/m^2),
L_i – luminance of glare source i in the direction of the observer's eye (cd/m^2),
ω_i – solid angle of the glare source
i seen from the observer's eye (sr),
P_i – position index (Guth's index) for the glare source
i according to Guth's analysis; this index is tabulated.

International standards define the rules for determining indexes and set out recommendations regarding the values for relevant working conditions using only three indexes. UGR for indoor workplaces [EN 12464-1], GR for outdoor workplaces (but also for indoors in specific environments) [EN 12464-2], and TI for road lighting [EN 13201]. Moreover, standardization work has not yet covered the glare from daylight. Nevertheless, considering the very large collection of publications on this problem, appropriate international standards should be expected in the near future.

It is worth noting that glare strongly affects the quality of lighting and well-being. And unfortunately, it is visible as a negative impact. We do not notice the problem – we are not aware of the quality of lighting – if glare does not occur. In contrast, work is difficult, we feel the working conditions are bad, and notice poor

FIGURE 4.2   The luminance maps of two workplaces (a) old luminaire is applied $UGR = 22$; (b) new luminaire $UGR = 18$.

lighting quality, when glare occurs. An example of the practical use of the UGR index for glare evaluation is shown in Figure 4.2. Glare evaluation was carried out in an industrial plant, in both cases (Figure 4.2 a and Figure 4.2 b) and in the same factory hall. Figure 4.2 a shows a luminance map for a workstation illuminated with old-type luminaires. The UGR index value calculated on this basis was 24. This value exceeds the limits ($UGR = 22$) for this workplace recommended in the standard. Figure 4.2 b shows a luminance map for the same workplace illuminated with a new type of luminaire. The measured value is $UGR = 18$, and fulfills the standard recommendation. The subjective assessment carried out by employees in both cases was consistent with the objective assessment. Descriptions of both luminaires can be found in Section 8.2.3.

There are many factors that affect glare and as a consequence influence the quality of lighting and well-being. Their impact should be considered in both lighting design and glare evaluation.

- Location of glare sources: studies carried out in the 1940s [Luckiesh et al. 1949] have shown that glare perception strongly depends on the location of the glare source in the observer's field of view. Guth proposed a parameter P (the so-called Guth index) [Luckiesh et al. 1949], which takes values from 1 (a position on the line of sight) to 15 (a position on the edge of the field of view). A mathematical relationship describing the value of the index as a function of position has been formulated, and the index values have been tabulated [CIE 1995]. The impact of the position index on discomfort glare is presented in Moosmann et al. [2009].
- The worker's age: in 1999, glare was found to have a stronger effect on the visual perception in the elderly [Higgins et al. 1999]. At the same time, older people show greater tolerance to glare [Wolska et al. 2014].
- Cultural and racial factors: in the 1990s, research was conducted [Pulpitlova et al. 1993] on a group of Japanese, Europeans, and Americans. Studies have shown that the Japanese are much more tolerant of discomfort glare

than the other groups, which may be due to cultural and anatomical differences (the shape of the eyelids, eye position in the eye socket).

- Time dependence related to glare: research was conducted on the speed of recovering visual acuity after glare. The authors, Irikura et al. [1999], found that increasing the illuminance of the glare source causes an increase in focus recovery time. Moreover, it has been experimentally shown that prolonged exposure time to glare also increases the recovery time [Lehnert 2001]. In addition, we can talk about a certain amount of glare (doze of glare) – the level of veiling luminance multiplied by duration time. The higher this dose, the longer the focus recovery time [Chen 2004]. At the same time, the focus recovery time decreases when the background luminance level increases [Irikura et al. 1999].
- Influence of color and spectral distribution of glare sources on discomfort: Bullough [2009] conducted experiments using light sources with different spectral distributions – from monochromatic to single color – of different spectra. He found that shorter wavelengths had a stronger effect on glare. In Bellia et al. [2011], the influence of spectral distribution on non-visual interaction was also analyzed.

The *UGR* is recommended by international standards to assess glare for indoor workplaces under artificial lighting. Analyses of the compliance of the measured discomfort glare index with human subjective assessment, supplemented with descriptions of basic measurement problems, are very widely presented in the literature [Akashi et al. 1996; Clear 2013; Sawicki et al. 2015; Wolska et al. 2013; Wolska 2013]. However, there are many works where it has been shown that assessment based on the index value is not always consistent with the subjective perceptions. As early as 1996, this problem was shown in research by Akashi et al. [1996]. In the work of Cai et al. [2013] research is presented, where the *UGR* value overestimated the level of discomfort glare compared to subjective assessment. There are many indications that research on the *UGR* will be continued and perhaps its verified form will appear. In addition, the application of new light sources, such as LEDs, means that completely new scientific challenges appear [Khan et al. 2015]. At the same time, the use of modern measuring methods expands the possibilities of research and physiological analysis of glare sensation mechanisms [Girard et al. 2019]. The use of fMRI [Bargary et al. 2015a] made it possible to analyze neuron response in specific cortical locations. The authors suggest that the sensation of discomfort glare is a response to saturation or hyperexcitability of visual neurons. The mathematical model of the retinal receptive field makes it possible to evaluate the glare level [Scheir et al. 2018; Donners et al. 2015].

The question of additivity of the glare phenomenon returns. Is it possible to analyze the feeling of glare using an appropriate index, and at the same time, will this index simply add up the luminance of the respective angular areas in the field of view? Of course, the direction of observation and the appropriate position index must be taken into account. Research to date for conventional light sources has confirmed this possibility and the usability of known indexes. Although specific glare conditions can be shown when simply adding up sources, this does not give adequate results [Sawicki et al. 2016b]. The study conducted on a group of 50 people [Bargary et al. 2015b] for LED with high luminance showed that discomfort glare in the central field of view in this case depends, above all, on local changes in luminance and contrast. The form of the contrast border is more important than the total amount of light entering the eye. Unfortunately, known glare indexes do not take such dependencies into account.

It should be expected that there would be no glare problem from well-designed lighting in the workplace. Unfortunately, this is not always possible. However, every effort should be made to minimize the glare phenomenon.

4.4 COLOR ASPECTS

The color of light is a subjective visual impression accompanying the process of seeing. We attribute color (as a property) to objects and light sources. From the point of view of physics, there is no such parameter (property) as color. We can only talk about wavelength and surface reflection properties. Color is just an impression produced in human consciousness based on the spectral analysis of radiation. On the other hand, color perception is very important for humans. It is one of the perceptive features that strongly influence well-being and the assessment of lighting quality.

Color, together with the feeling of brightness, characterizes the perception of visual sensations. The recognition of contrasts (Section 4.2.3) is necessary to interpret an image, giving the ability to distinguish between details and shapes of objects. In many situations it is necessary, but not sufficient. Only the correct recognition of color gives the full interpretation of the perceived image. Color impression is created in our brain based on nerve signals coming from the visual system (see Chapter 2). Humans are used to living in a natural environment illuminated by sunlight. The sun is a source of light with a continuous energy spectrum in the visible area of electromagnetic radiation. Artificial lighting resulting from the combustion of various materials is also characterized by a continuous spectrum. Although the climatic conditions and the rhythm of daily and annual changes affect the color experience associated with sunlight, our experience allows us to see colors as expected. A red tomato is seen to be red, regardless of the time of day or year. It is affected by neither moonlight nor candlelight. Humans can adapt color impressions to their experience – this mechanism is called chromatic adaptation. Chromatic adaptation allows us to correctly interpret color sensations in different environmental conditions, when the light reflected from the surface of the object comes from sources with different spectral compositions. A photo taken in such conditions with a camera requires correction – white balance, so that the colors in the photo are correct.

The CIE XYZ (CIE 1931) color space developed in 1931 by CIE [Smith et al. 1931–1932] is the basic space used in colorimetry. Figure 4.3 shows the chromaticity diagram for CIE XYZ. The black body temperature emission line (the so-called Planckian locus) is marked in the drawing. The points on this curve correspond to the color coordinates for black body radiation at the appropriate temperature. It is difficult to clearly define white for natural solar lighting. This is caused by changes (daily and yearly) in the color temperature of solar radiation – meaning that under different conditions, solar radiation treated as white will have a different spectral

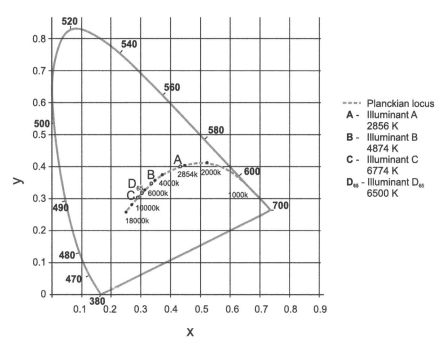

FIGURE 4.3 Chromaticity diagram of the CIE XYZ (CIE 1931) color space.

composition. In this way, objects illuminated by this radiation can give different color sensations. In order to compare (and also copy and reproduce) colors in such conditions, a set of standard color distributions of the so-called illuminants was introduced. These colors correspond to the color commonly recognized as white, but created under strictly specified conditions.

- Illuminant A, color temperature 2856 K, represents an incandescent tungsten filament lamp.
- Illuminant B, color temperature 4874 K – average daylight in direct sunlight.
- Illuminant C, color temperature 6774 K – shady (diffused) daylight.
- Illuminant series D_{TC} – daytime radiation for different times of the day and appropriate color temperatures, e.g. D_{65} for color temperature 6500 K.

A set of chromaticity coordinates supplemented with a specific illuminant at which the color is analyzed clearly defines the color of the light.

4.4.1 CORRELATED COLOR TEMPERATURE (CCT) AND ILLUMINANCE LEVEL

The correlated color temperature (CCT) of the light that accompanies our activities strongly affects our emotions and well-being. People prefer warm colors – most often associated with safety and peace. Studies show that changing the level of light intensity affects the color impression: light with high color temperature at low intensity is perceived as cool and unpleasant, light with low color temperature at high intensity is

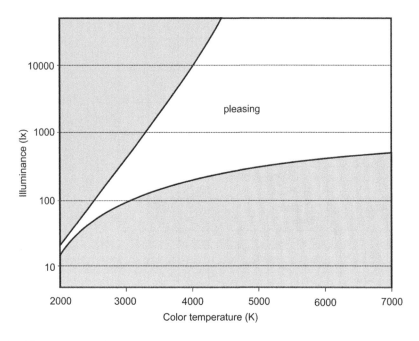

FIGURE 4.4   Kruithof diagram.

perceived as unnaturally saturated. The Kruithof diagram [Kruithof 1941] showing the acceptability of colors depending on the correlated color temperature and illuminance level is well known (Figure 4.4). When designing artificial lighting, choose the right combination of correlated color temperature (CCT) and illuminance level. Thanks to this, the luminous environment will be perceived as pleasant. Generally, it can be stated that at higher CCT, higher illuminance should be used.

The Kruithof diagram has been repeatedly analyzed in various studies. Modern research with particular emphasis on LED [Viénot et al. 2009] indicates the need to correct the diagram. The result of these analyses was a proposal of recommendations presented in the CIE Technical Report [CIE 2014a]. The article [Fotios 2017] provides an overview of 29 different studies. Nine of them were considered the most important and were thoroughly analyzed. After comparing their results, the authors formulated their conclusions. The results show that the impact of color temperature (CCT) on subjective ratings of pleasantness and brightness is very small. In addition, an increase in illuminance from 300 lx to 500 lx improves pleasantness, but a further increase above 500 lx affects feelings to a minimal degree. On the other hand, the feeling of security, warmth, and pleasure associated with lower CCT levels is common. Should we therefore discard the Kruithof diagram? It seems it is too early for such a radical step [van Bommel 2019]. The Kruithof diagram shows the relationships for a particular combination of parameters, which are, in addition, often consistent with common feelings. However, it may not be valid for general use. Moreover, analysis of the problem shows that many other aspects should be taken into account [Cuttle 2017], including such personal characteristics as mood, attitude to life, expectations, and chronotype. The problem is certainly very important and interesting, so research will continue.

There are ongoing studies concerning the relationship between CCT, illuminance level, and comfort, especially for LED. One of the latest works in this area has been devoted to psychological and physiological responses [Kakitsuba 2020] of subjects (women and men) exposed (during the daytime between 1 pm and 5 pm) to light emitted from LEDs of 3000 K, 4000 K, and 5000 K and two illuminance ranges: low illuminance range (between 70 lx and 400 lx, depending on CCT) and high illuminance range (between 1500 lx and 7000 lx, depending on CCT). The authors stated that gender does not affect the assessment of boundary illuminances, based on psychological and physiological responses associated with the combination of different CCTs; however, the high boundaries of LEDs were higher than those of fluorescent lamps at all CCTs tested. Nevertheless, these differences may be due to different spectral distributions and luminances for both types of light sources.

Unfortunately, a new diagram for the selection of illuminance and CCT (which would replace the Kruithof diagram) has not yet been proposed, but research will make it possible to specify the new rules for this selection.

4.4.2 Color Perception

A very important mechanism in the discussion of colors is their perception, independent of their spectral composition. This phenomenon is called metamerism. Metamerism is the term used to describe the fact that two light sources of different spectral power distribution may produce the same perception of color (illuminant metamerism) or that the same light sources may produce different perceptions of color in different observers (observer metamerism) [Setchell 2012]. According to Grassman's third law, lights of the same color give the same impression when mixed with other lights, regardless of their spectral distribution. The color of the mixture depends only on the colors of the ingredients, but not on their spectral distribution. A particular color spectrum corresponds to exactly one color. However, a given color can correspond to infinitely many different spectral distributions of radiation.

From the point of view of color perception, it is also important for a person to perceive the color of objects in the same way under different lighting. This is not only crucial for well-being, but especially for work performance. It allows people to correctly distinguish objects and correctly perform visual tasks. This means the need to provide the same (similar) perceptual impressions, with different spectral properties of artificial light sources. The issue is known as color rendering, and has been particularly rapidly developed in recent years, primarily due to the introduction of new light sources – LEDs. In the context of color rendering, both standards (EN 12464-1 European standard [2011] and WELL Building Standard [WELL 2019a]) are limited to describing the need to ensure correct color perception. As for color rendering indexes, only R_a (and R_9) is discussed in WELL. The subject of color rendering as a problem related mostly to visual performance and light sources (primarily LEDs) is discussed in Section 5.8.

The phenomenon of metamerism has always been associated only with the visual effect. However, research also shows the relationship between this phenomenon and the non-visual effect. The authors of the work by Aderneuer et al. [2019] presented the possibility of tuning circadian effects using LEDs. They used sources emitting white light that appear similar, but have different spectral distribution.

4.5 PERCEPTION OF FLICKER AND THE STROBOSCOPIC EFFECT

The world we live in is dynamic. Proper perception of the movement of objects is very important to us and especially if we consider ensuring safety. Lighting should not be changed in its intensity and color too quickly, because this negatively affects the human nervous system, disrupts correct perception, and causes discomfort and fatigue. Lighting should appear unchanged over a short timeframe. We are used to natural sunlight, which, although it changes in annual and daily cycles, ensures stability over the short periods of time needed to carry out visual tasks. It is different from artificial lighting. Various modern light sources emit light that changes over time. There are two basic phenomena associated with cyclical changes in the intensity of light emitted by artificial light sources: flicker and stroboscopic effect.

Both standards (EN 12464-1 European standard [2011] and WELL Building Standard [WELL 2019a]) treat the problems of flicker and the stroboscopic effect briefly. However, the considerations come down to recommendations for limiting flicker. The properties and parameters describing the phenomenon are not analyzed. This is because flicker is most often associated with efficiency and work performance, and above all with its hazardous effects and safety at work [van Bommel 2019]. The assessment of flicker (and the stroboscopic effect) is difficult. The basic parameter associated with flicker is the frequency of changes (flicker frequency – ff). However, analyzing the phenomenon requires additional parameters and is quite complicated. It also depends on the features typically associated with visual performance. The problems of flicker perception and the stroboscopic effect are discussed in the context of visual performance in Section 5.7. The parameters needed to measure and assess the phenomena are examined in Section 10.2.

4.6 SUMMARY

The quality of lighting is crucial for the visual impact of light on humans. Ensuring visual comfort requires taking into account a number of parameters related to the intensity of light, its spatial distribution (luminance distribution, discomfort glare) and the aspects of color which are described in this chapter. Visual comfort and visual fatigue related to lighting quality influence the well-being of workers, also including their emotional and physical health.

Although standards provide lighting requirements the fulfillment of which is to ensure adequate lighting quality, the use of the recently introduced LEDs for lighting purposes means that some of the lighting parameters are inappropriate for assessing the quality of such lighting. This applies to color rendering, for example. Therefore, new color rendering indexes dedicated to the adequate assessment of LEDs have been developed in recent years. For this reason, it is recommended to use R_f and R_g instead of R_a when assessing the color rendering properties for LED lighting.

5 Visual Performance

In the right light, at the right time, everything is extraordinary.

Aaron Rose

5.1 VISUAL PERFORMANCE: DEFINITION AND RECOMMENDATIONS

Visual performance is usually defined in terms of the speed and accuracy of processing visual information. In order to evaluate visual performance for particular lighting conditions, it is necessary to have measurable human responses (e.g. visual response time, data on the number of mistakes).

One of the important aspects of good lighting is to create lighting conditions which favor high visual performance. To achieve this, knowledge of the human visual system must be implemented in order to adapt lighting to human capabilities.

The first recommendations for indoor lighting which took into account visual performance and human visual capabilities were developed by the CIE in the form of the 1981 publications [CIE 1981a; CIE 1981b]. CIE recommendations were based on Blackwell's [1981] studies conducted on large age-diverse groups. Blackwell also showed that with age we experience considerable loss of the ability to distinguish contrasts. However, the tests that were analyzed were carried out in strictly defined conditions and only for a selected visual task. Later studies attempted to consider additional factors related to the type of task, lighting, and the environment in which the task is being carried out. Unfortunately, attempts to make generalizations did not bring about the desired results and the project was not continued.

5.2 WEBER'S FRACTION AND CONTRAST

We expect lighting in the workplace to create conditions for specific visual tasks. The most important visual feature in the task field is the difference in luminance and/or color. This difference is usually called "contrast", and is determined on the basis of changes in the luminance or colors of perceived objects. Contrast of the detail of an object with its background is a prerequisite for vision in general. If contrast is too low, we are unable to detect and determine the shape of the object or its movement. Of course, detection alone is not enough for work; identification is necessary, which usually requires even higher contrast than detection.

5.2.1 THE WEBER FRACTION

Luminance changes which occur in the field of view can be characterized by the relationship between the luminance values of bright and dark places. The ratio of the local luminance increment to ambient luminance that occurs at a specific location determines the local contrast. This increase is called Weber's contrast or the Weber fraction [Whittle 1994]. If the object with luminance of L_a is observed on a background with luminance of L_b, the contrast determined by the Weber fraction (WF) is described by Formula (5.1).

$$WF = \frac{L_a - L_b}{L_b} \tag{5.1}$$

It is worth noting that the Weber fraction defined in this way can have positive values (light objects on a dark background), as well as negative values (dark objects on a light background).

The smallest value of the Weber fraction for objects distinguished by the eye is about 1% in conditions of uniform luminance distribution [Wyszecki et al. 1982]. In conditions of non-uniform luminance distribution, Weber's fraction increases to 10%. At the same time, this value strongly depends on individual human properties and the level of luminance in the field of view. It is assumed that this value ranges from 0.02% to 2% (when we consider uniform luminance distribution).

Weber's law says that the value of the Weber fraction should be constant, regardless of the luminance value. In fact, the dependence of the Weber fraction on luminance is not linear, as presented in Figure 5.1.

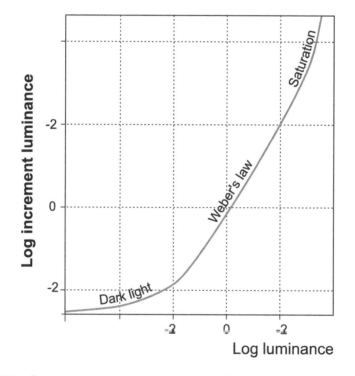

FIGURE 5.1 Log increment luminance as a function of log luminance.

We can distinguish the luminance range for which the Weber fraction is constant – with a large approximation. It is assumed that this applies to the range from approx. 50 cd/m^2 to approx. 10,000 cd/m^2. Outside this scope, we can say that Weber's law does not apply.

The Weber fraction represents the threshold of luminance contrast – the just noticeable difference (*JND*), below which luminance changes are not detected by the human eye (in other words, this is the smallest perceived luminance difference). The human eye is relatively less sensitive to changes at low luminance levels and more sensitive to changes at high luminance values. This means that it is easier for a worker to recognize small details in bright than in dark illumination of the field of view.

5.2.2 LUMINANCE CONTRAST

Luminance contrast could be defined by several formulas. The often-used equation of contrast *C*, Formula (5.2), is applied especially for objects that are brighter than their background [Rea 2000]:

$$C = \left| \frac{L_d - L_b}{L_b} \right| \tag{5.2}$$

L_d – luminance of object, L_b – background luminance.

Another formula of contrast (or modulation) *CM* was developed by Michelson [1927] (Formula (5.3)). It is used for analyzing cases of multiple, periodic luminance changes (from maximum to minimum) of the object observed. This applies to gratings which have one maximum and one minimum in each cycle [Rea 2000] or images with a string of alternating black and white lines on the monitor screen.

$$CM = \frac{L_{max} - L_{min}}{L_{max} + L_{min}} \tag{5.3}$$

Studies of contrast modulation perception were most often carried out using a grating (sinusoidal) or checkerboard form [Kolb et al. 1995; Van Bommel 2019].

Let us draw attention to the fact that in a real workplace, ideally balanced visual conditions, where the visual task is surrounded by surfaces of uniform luminance, are practically non-existent. There are many different objects in the observer's (worker's) field of view and the surrounding surfaces may have a complex finish (different colors, direct and diffuse reflections etc.). In this case, it can be difficult to distinguish between adjacent objects or define the boundaries (edges) of the object. In addition, it should be mentioned that human perceptive capabilities depend on changes in spatial frequency.

5.3 CONTRAST SENSITIVITY

5.3.1 CONTRAST SENSITIVITY FUNCTION

The eye exhibits variable sensitivity to contrast, depending on the spatial frequency of image fragments. The spatial frequency of the image is expressed by the number

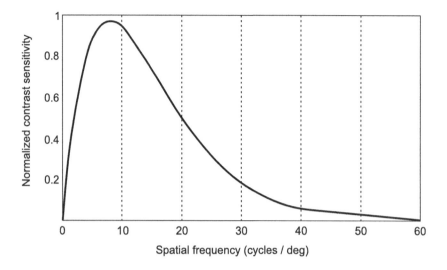

FIGURE 5.2   Normalized *CSF* as a function of spatial frequency, based on Formula (5.4) from Mannos [1974].

of pairs of parts (lines, bars etc.) of contrasting luminances (light and dark) producing an image within a given distance on the retina. The unit of retinal distance is usually one-third of a millimeter. This is convenient, because an image of such a size subtends one degree of the visual angle on the retina. Spatial frequency is expressed as the number of cycles per degree of visual angle. When there are many objects/details in the field of view (located close to each other) that differ in luminance, we are talking about high spatial frequency. For large areas with uniform luminance, the term low spatial frequency is used. The relationship between contrast perception and spatial frequency of the image is described by *CSF* – contrast sensitivity function [Bovik 2009]. Various studies indicate that the maximum contrast sensitivity of the human visual system lies in the range of spatial frequency between 5 and 10 cycles per degree [Schieber 1992], which is shown in Figure 5.2. Maximum contrast sensitivity depends on the viewing angle. When the angle decreases, the maximum value of contrast sensitivity shifts slightly towards larger values of spatial frequencies. The inverse relationship applies to the luminance of the observed image: with a lower luminance, the maximum value of contrast sensitivity moves towards lower values of spatial frequency. All studies clearly confirm that both lower and higher frequencies are gradually suppressed in the human visual system. The reduction in sensitivity with increasing spatial frequency is explained by the finite possibilities of the eye as an optical system. On the other hand, contrast suppression at lower spatial frequencies is considered a specific psychophysical phenomenon. Attenuation in the lower spatial frequency range is associated with poor vision of large objects, which can lead to problems with movement in unfamiliar surroundings. Attenuation at higher spatial frequencies is associated with the recognition of details (and relatively small objects), for example, it can lead to difficulties in facial recognition. It is worth noting that human perception capabilities in perceiving details deteriorate with age, as a result of which what is impaired in older people is mainly recognition of higher spatial frequencies.

Mannos introduced the concept of normalized CSF (CSF_N) for gray scale images as a function of spatial frequency (f_s) – Formula (5.4). The relevant graph is presented in Figure 5.1 [Mannos and Sakrison 1974]. The normalized CSF function is a generalization of the function for various operating and observing conditions (luminance of objects, viewing angle, observation distance). The real CSF function will always have a similar shape, but may differ in contrast sensitivity values under specific conditions. The normalized CSF function reaches the maximum value for 8 cycles/degree.

$$CSF_N = 2.6(0.0192 + 0.114f_s)e^{-(0.114f_s)^{1.1}} \qquad (5.4)$$

where $f_s = \sqrt{f_h^2 + f_v^2}$

f_h, f_v – horizontal and vertical frequencies.

In addition, people evaluate contrast differently from various viewing directions in the visual field. Humans are able to distinguish contrast relatively better horizontally and vertically than diagonally [Campbell 1966]. At about 60 cycles per degree, contrast sensitivity drops practically to zero – this level has been confirmed by many different studies [Westheimer 1964; Hoekstra et al. 1974]. However, this is the maximum spatial frequency value (60 cycles per degree) occurring when luminance is at the level of 300 cd/m^2 [Boyce 2014]. Reducing the luminance of the objects reduces the perceived spatial frequency. The graph shape of contrast sensitivity vs spatial frequency depends on luminance and will be different for each luminance value. It is important, however, that all graphs are below (within the bounding box) the maximum contrast sensitivity graph for optimal lighting conditions. The human eye is able to adapt to changing lighting conditions and to both very high and low levels of luminance. For each luminance level, the graph will always be of a band-pass filtering type, but maximum contrast sensitivity is for different spatial frequency values (Figure 5.3).

5.3.2 CONTRAST SENSITIVITY AND SUPRATHRESHOLD VISIBILITY

For complex images (many spots with different levels of luminance and many elements with different spatial frequencies), adaptation luminance changes locally. An attempt to estimate the value of local adaptation for a complex image comes down to decomposing the image into a number of sub-images and conducting a local analysis. When adapting to the level of lighting in the field of view, we are talking about suprathreshold visibility. Changing the adaptation luminance affects both the perception of details and the perception of colors. Analysis of suprathreshold visibility was carried out in the 1960s [Stevens 1961; Stevens et al. 1963]. In the suprathreshold visibility range, depending on the contrast between the object and the background, an increase in the brightness of the background causes an increase or decrease in the brightness of the object. As brightness increases, saturation increases. As a result, bright objects can be seen as brighter and dark objects as darker. This is not in accordance with Weber's law.

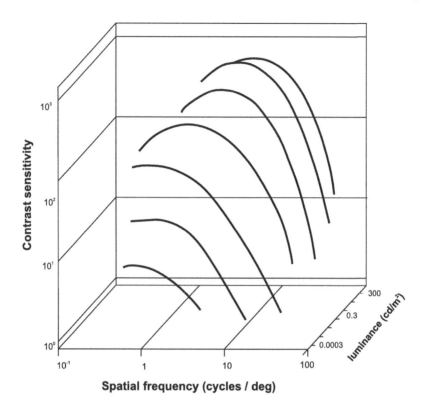

FIGURE 5.3 *CSF* dependence on spatial frequency for different levels of luminance, based on data from van Nes [1967].

Apart from low spatial frequencies, contrast sensitivity increases with increasing stimulus luminance. However, above a certain value of luminance level, increasing it further not only does not increase the perceived spatial frequency, but becomes disturbing to humans. The Weber fraction function is linear (with a logarithmic scale of light stimulation) only for a certain range. This means that in order to obtain optimal conditions for distinguishing objects (details in the workplace), an adequate illuminance level should be provided depending on the working conditions. The nature of the changes in the Weber fraction is consistent with the overall response curve of human senses to stimulation (Figure 5.4). It has an S-shape form and describes this type of relationship for all the senses [Lipetz 1971]. It is often assumed that for an object to be easily visible, it is necessary that the luminance contrast is at least twice the contrast threshold. Contrast analysis is performed for different levels of luminance without taking into account the spectral distribution of light. It has not been proven that with the same luminance of white light sources, the function of contrast sensitivity depends on the spectral distribution of light.

The contrast at which the accuracy of perception is 50% is called a threshold luminance contrast [Rea 2000]. (Threshold luminance contrast is defined as a level of stimulation which induces accuracy of perception at the level of 50%). The cited parameters of sight properties are based on tests on large groups of

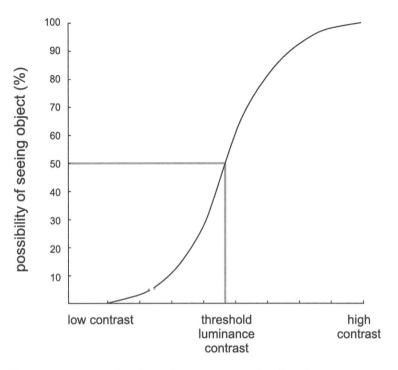

FIGURE 5.4   S-shape function. Perception accuracy as a function of contrast.

people and appropriate statistical analyses. There may be individual differences among workers. The level of contrast at absolute threshold, i.e. the ability to perceive details, depends on many factors: individual predispositions, age, employee experience, as well as the size and shape of the detail, and the time of observation. Let us also remember that in addition it depends on the worker's level of emotions and stress.

5.4   VISUAL ACUITY

Visual acuity is the factor that strongly depends on the human visual system. Visual acuity refers to the human ability to discern the shapes and details of the optotypes presented. Considering the structure of the retina, one can speak of the perception of two objects as separate when their images recorded by the cones on the retina are separated by at least one cone. Assuming that the cone diameter is about 0.004 mm, visual acuity is about 1 angular minute. It is worth noting that this corresponds to the perception of contrast in the *CSF* function, where the absolute limit is 60 cycles per degree. Visual acuity is also affected by the illumination level on the detail/object observed. The relationship between background illuminance of the target observed and visual acuity was determined on the basis of observation studies of Landolt rings (Figure 5.5). For $S = B$, the higher the background luminance, the better the visual acuity. For S at the very low level, there exists an optimal level of B (the best acuity). Increasing B above this level decreases acuity.

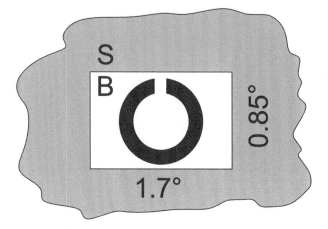

FIGURE 5.5 Black Landolt ring in a background field (0.85° by 1.7°). B – luminance of the background, S – luminance of the surround field. Based on Boyce [2014].

The optimal range of the level of illuminating the visual task can be determined by considering the perception-related properties of contrast discussed here. The minimum level is one that provides object discrimination at maximum visual acuity. The maximum lighting level is one at which the Weber fraction maintains a constant value.

5.5 RELATIVE VISUAL PERFORMANCE (*RVP*)

Due to the complexity of the problem, it is probably not possible to make illuminance recommendations (as a functional dependency) based on visual performance and visual comfort [Boyce 1996]. This means that it is not possible to choose lighting conditions to meet everyone's expectations and take into account all the important parameters. Nevertheless, since we are looking for practical solutions for lighting applications, modeling attempts to design lighting were developed. A model which is widely recognized today as the best solution is based on relative visual performance [Rea et al. 1991]. It took many years to develop this solution [Rea 1981; Rea 1986; Rea et al. 1988] and it is based on earlier works [Weston 1953; Weston 1961] in which the Landolt ring was used.

In the *RVP* model, relative visual performance is shown as a non-linear function of task parameters. Here we can discuss such parameters as contrast, angular size of object, lighting level (in the original papers these are referred to in trolands, but they can be represented by illuminance), and also the age of the employee. The non-linear function is represented in the shape of a plateau and escarpment of visual performance. The *RVP* model was later extended by adding the parameter of reaction time. It is worth adding that the *RVP* model has been independently validated [Eklund et al. 2001].

Conclusions from the research conducted on the *RVP* model can be reduced to several important recommendations. Generally, the increase of illuminance and increase of contrast have an important impact on improving visual performance.

However, if we consider visual tasks involving objects of large size and the contrast is high, it is not worth increasing the light intensity too much, because we will achieve saturation of work efficiency relatively quickly (reaching a lighting level at which a further increase in intensity does not improve performance). On the other hand, two tasks can be considered: the first one with small objects and low contrast and the other one with large objects and high contrast. In practice, we never have the possibility of equalizing the level of performance of these tasks by increasing the illuminance in the first of them. If it is possible to make an improvement in contrast and/or increase the size of the objects, then such changes should be preferred. This will always bring better results than just increasing the illuminance.

The *RVP* model of visual performance also has limitations. This model was prepared for achromatic visual tasks observed in photopic conditions. If there are color differences, the luminance contrast can tend to zero. This way, visual performance will be maintained. In the *RVP* model, on-axis visual tasks are discussed. For off-axis tasks, the trends will be the same, but the changes (*RVP* level) will be stronger. For mesopic conditions, the effect of luminance could be different depending on spectral properties, especially for off-axis tasks and tasks corresponding to peripheral retina perception. What has been included in the *RVP* model is visual performance, not task performance. This is important, because sometimes non-visual performance can have a significant impact on the task; however, it will not be included in the model under consideration. The *RVP* model is considered to work properly for a certain class of tasks – hence the conclusion that work on this model is worth continuing [Boyce 2014]. Examples of *RVP* relationships for various parameters in the *RVP* model are presented in the charts in Figure 5.6.

5.6   THE LATERAL INHIBITION PHENOMENON

Perception of details and distinguishing between objects is facilitated by the phenomenon of lateral inhibition. The phenomenon of lateral inhibition occurs on the retina of the human eye [Bakshi and Ghosh 2017]. Retinal neurons which are stimulated affect the neighboring neurons; moreover, a neuron's response depends not only on the level of luminance (color) observed, but also on the luminance (color) of the neighboring areas. The effect of stimulated neighboring neurons (photoreceptors) makes the eye more sensitive to local changes in luminance than it would appear from its real level. This way, local changes are perceptually increased [Purves et al. 2003; Snowden et al. 2012] (Figure 5.7). Studies show that the phenomenon is not dependent on color; lateral inhibition occurs in the same manner, regardless of the spectral properties. Lateral inhibition is very important in the human visual system. It allows us to recognize details in low light, e.g. we can read a book in moonlight. The phenomenon is beyond human control. It is not possible to deliberately use it when illuminating the workplace. However, in most cases, regardless of the environmental conditions (e.g. those considered in the *RVP* model), it helps people to perform their visual task.

However, examples can be shown in which lateral inhibition can interfere with the perception of visual experience. Modern light sources and electronic control options allow us to create lighting properties almost freely. For example, light spots

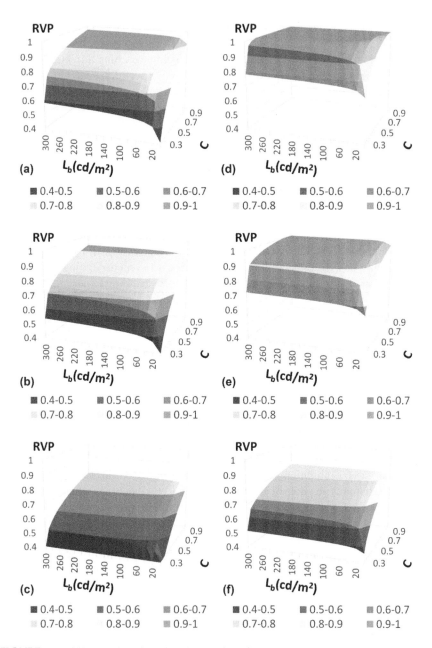

FIGURE 5.6 *RVP* as a function of background luminance (L_b) and object contrast (*C*). Two visual angles have been taken into account: 4.1' and 8.8' (as 4-point and 8.5-point Times New Roman letters viewed from 50 cm [van Bommel 2019]. Age: (a) 20 yrs, (b) 40 yrs, (c) 60 yrs, (d) 20 yrs, (e) 40 yrs, (f) 60 yrs. Graphs based on formulas describing Weston's data given in van Bommel [2019].

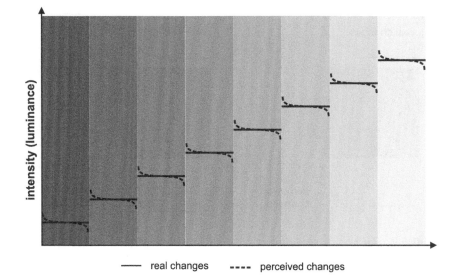

— real changes   ---- perceived changes

FIGURE 5.7   The phenomenon of lateral inhibition.

of different luminance or color are used, or such spots are mixed. Note the example of mixing two colors of different luminance. It would seem that if we want to achieve a linearly perceived change in color, then the level of luminance should be changed linearly. Unfortunately, such a change brings about very bad results. The phenomenon of lateral inhibition will disrupt perception (Figure 5.8 a). Studies described in [Sawicki et al. 2019] show that in this case, a non-linear change in luminance will be most appropriate (Figure 5.8 b). A good solution is to use Penner's easing functions [Penner 2002; Izdebski and Sawicki 2016].

5.7   FLICKER AND STROBOSCOPIC EFFECT AS A HAZARDOUS PROBLEM OF PERFORMANCE

The world we live in is dynamic. Proper perception of the movement of objects is very important to us and our safety. Lighting should not be changed in its intensity and color over short periods of time, because it negatively affects the human nervous system, disrupts correct perception, and causes discomfort and fatigue. Over short timespans, it should appear unchanged. We are used to natural sunlight, which, although it changes in annual and daily cycles, ensures stability over the short periods of time needed to carry out visual tasks. This is different with artificial lighting; however, there are various modern light sources which emit light that changes over time. There are two basic phenomena associated with cyclical changes in the intensity of light emitted by artificial light sources: flicker and stroboscopic effect.

Flicker is an impression of unsteadiness of visual perception induced by a light stimulus whose luminance or spectral distribution fluctuates with time [CIE 2016a]. The dependence of contrast sensitivity on changes in temporal frequency can be described as similar to the dependence of the contrast sensitivity function (*CSF*)

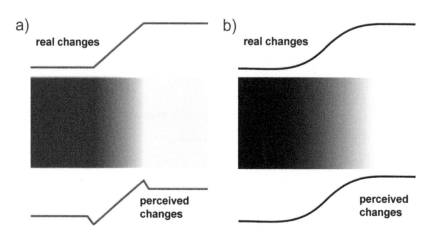

FIGURE 5.8 (a) linear change – artifacts appear as a result of lateral inhibition; (b) example of using Penner's function – an impression of a smooth change in brightness arises.

on spatial frequency (Figure 5.3). In the literature describing these types of graphs, visible flicker sensitivity (*VFS*) (Figure 5.9, temporal modulation transfer function (*TMF*), or temporal contrast sensitivity function (*TSF*) are often given instead of *CSF* [Boyce 2014; Hart 1987; Rea 2000]. All of these parameters correspond to the inverse of *CSF*.

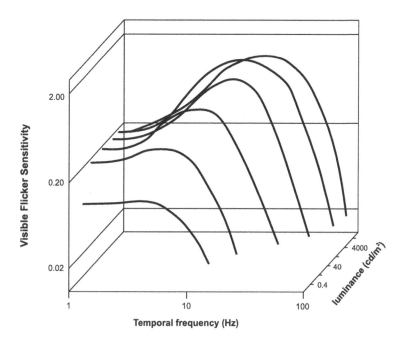

FIGURE 5.9 Visible flicker sensitivity (*VFS*) as a function of temporal frequency and luminance. Based on data from Kelly [1961].

Increasing luminance increases flicker perception. The contrast detection threshold increases with luminance to approx. 300–450 cd/m^2, where flicker sensitivity is highest at flicker of 10–20 Hz. Above this maximum level, an increase of luminance does not raise the maximum contrast detection threshold. In this case, the frequency range of flicker detection increases, but this change is small. The highest frequency perceived as flicker in optimal conditions and at maximum contrast is about 80 Hz. Above this frequency, flicker is not perceived. There is an effect of persistence of vision phenomenon (retinal persistence) [Kolb et al. 1995], in which the impression of an object image persists in the brain through neurons from the retina for about 1/20 s after the light stimulus. This phenomenon makes it possible to produce films, and also explains why the ability to detect contrast decreases with increasing flicker frequency.

5.7.1 CRITICAL FUSION FREQUENCY

The perception of flicker frequency also depends on the size of the object that changes, i.e. the size the object occupies in the field of view.

The maximum perceived flicker frequency is called critical fusion frequency (CFF). Above this frequency we perceive the flickering light as continuous. The value of CFF depends on many factors [Kolb et al. 1995].

- Luminance of light source. The higher the luminance value, the higher the CFF value. In a wide range of light source luminance, CFF increases in proportion to the log luminance: Ferry–Porter law.
- Eye/retinal position. The value of CFF is different when perceived by rods and cones, so CFF is different for central (paracentral) vision and peripheral vision.
- Spectral dependencies. Under photopic vision, the effect of source luminance does not depend on the spectrum of the source. Under scotopic vision, the effect of luminance depends on the spectrum of the source.

With well-designed lighting in the workplace, we expect that flicker will not be perceived and will not affect the health of employees. Flickering light is always seen as a disturbing factor. Under certain working conditions flicker can be a hazard to health. The phenomenon can be eliminated by providing such a level of adaptation luminance that flicker is not perceived in the field of view. However, under real working conditions it can be practically impossible. An effective solution may be to increase the flicker frequency of light sources above the level of perception. Different electronic devices are used for this purpose, such as electronic ballasts, drivers, and power supply units. For fluorescent and high intensity discharge lamps, for example, flicker can be minimized by using high frequency electronic ballasts.

It is worth paying attention to the color relationships in detecting contrasts. The visual system processes chromatic (color) contrasts and achromatic (luminance) contrasts on separate nerve pathways. This includes changes in both space and time. Image blur strongly reduces sensitivity to achromatic contrast, but sensitivity to chromatic contrast is not reduced. For objects larger than 0.5°, chromatic contrast is more noticeable than achromatic contrast, even if the image is blurred. On the other hand, differences between green and red objects are more noticeable than between yellow and blue ones. Sensitivity to chromatic contrast and sensitivity to achromatic contrast are not related to each other. This means that for low frequencies (0.5–3 Hz) it is easier to notice chromatic (color) changes, while for higher frequencies (about 10 Hz) changes in achromatic (luminance) variation are noticed faster.

The stroboscopic effect influences the perceived motion of rotating or reciprocating machinery [EN 12464-1]. It occurs when we observe moving objects illuminated by flickering light. The coincidence of flicker frequency and motion frequency (e.g. of rotating machine components) can lead to erroneous perceptions of motion. For example, a moving object can be seen as stationary. In this phenomenon, motion is interpreted by the brain as successive images and stitches them together for temporal continuity. Stroboscopic control of repetitive motion, such as the rotation of a wheel, can create an optical illusion that is completely contrary to the true motion.

The stroboscopic effect can also be felt for flickering light sources exceeding 80 Hz. The latest research shows that a person can feel this phenomenon for the flicker frequency in the range from 40 Hz up to about 1000 Hz [van Bommel 2019]. But we can notice the peak of the sensitivity at around 75 Hz for sinusoidally modulated light and at 250 Hz for square waveform modulation [van Bommel 2019].

5.8   COLOR RENDERING

5.8.1   ARTIFICIAL LIGHT SOURCES

Artificial light sources can be divided into two groups. The first group comprises phenomena in which radiation has a continuous spectrum, e.g. a light bulb. The second group consists of light sources generating a discontinuous spectrum, e.g. fluorescent lamps and LEDs. Spectral discontinuity means that within certain ranges of optical radiation, the source emits no light, or this emission is residual. Of course, knowing the properties of such a source, radiation emission ranges can be selected in such a way as to give the impression of emitted white light. The phenomenon of color additivity (one of the basic principles of colorimetry) provides this possibility. However, the impression of emitted white light is not enough for the correct perception of the colors of objects illuminated by such a source. The lack of emissions in certain bands makes the light reflected from the surface of the object acquire a different color than that to which a person is accustomed, i.e. that which is perceived when this object is illuminated by natural light (sunlight). The human environment in which objects have unnatural colors may be associated with misinterpretation of how objects appear. This can lead to problems with the assessment of shapes, details, sizes, or distances, so that the appearance of the environment or selected objects and their properties is interpreted incorrectly.

What we expect from lighting in the field of color rendition is that the color impression resulting from the use of this lighting should be in line with human expectations (and habits). If there are slight differences, they cannot lead to interpretation problems. For comparison of lighting properties in the range of color rendition, natural (solar) lighting is taken as the reference. Formally, color rendering of a light source is defined as an effect of an illuminant on the color appearance of objects by conscious or subconscious comparison with their color appearance under a reference illuminant [CIE 2016a].

Minor, acceptable differences in color rendering mean, for example, that products in the grocery store will still look natural and as expected, although objectively there will be differences in the spectral properties of colors.

5.8.2   Color Space

From the mathematical point of view, color is described by coordinates in the appropriate color space. The description of color is based on the three-component color theory. Dozens of different color spaces are known [Kang 2006; Mokrzycki et al. 2011]. Most of them are used in practice. This is due to technological or perceptual conditions. On the other hand, all the spaces used are derived from the CIE XYZ space. Although this space is uneven, the tradition of its use means that all comparative analyses are still carried out in relation to the CIE XYZ space. Humans are able to distinguish only a specific set of different colors. We can talk about minimal differences in the color coordinates which cause perceptibly different color impressions. An attempt at a mathematical description of perception in this area requires defining the operation of the color comparison in the appropriate space [Mokrzycki et al. 2011].

Because different light sources emit light with different spectral distributions (continuous or discontinuous), we would like to be able to assess what the objects illuminated by these sources look like. By analyzing the differences in the color perception of illuminated objects, a variety of sources can be compared. The assessment of colors could also be carried out subjectively, but a simpler solution is to compare the coordinates in the appropriate color space, knowing how these differences are perceived by humans. In this way, comparing a given source with a reference source for humans (natural – solar), one could assess how far the color of an object illuminated by a given source differs from the color of an object illuminated by the reference source.

5.8.3   Color Rendering Index R_a

The most commonly used index is the CIE Color Rendering Index (denoted as *CRI* or R_a) described by CIE in their 1995 publication [CIE 1995a]. It is worth remembering, however, that it was defined as early as 1964 and amended twice – in 1965 and 1974. The CIE method is based on a comparison of a set of color test samples, which are illuminated by a test source and the reference source. Among the 14 samples developed, eight have been prepared with moderate chroma (saturation), similar (approximately the same) lightness, and different hue. The next four special color samples are recommended by CIE for more specific evaluations. A typical source comparison includes the first eight samples only. As a reference source, an incandescent (Planck) source is used to test the sources with CCT below 5000 K. For sources with higher CCT, an appropriate mathematical model describing the spectral distribution of daylight is used. The R_a index is determined on the basis of coordinate differences specified for the samples compared. Calculations are carried out in uniform UV*W* space [Boyce 2014]. R_a takes a value less than 100. It is assumed that a source with $R_a = 100$ corresponds to the quality of the incandescent or natural solar source. The smaller the R_a value, the worse the color rendition property of the light source. There is no official lowest limit of value for R_a; however, the lowest value equal to 20 can be found in lighting standard requirements [EN 12464-1], but with an additional requirement that safety colors shall be recognizable. There are light sources that not only do not show samples in their real colors, but change (falsify) their colors. In such cases, R_a can even take negative values.

The advantages of this method primarily include the simplicity of testing and a well-chosen collection of test samples. Nevertheless, the disadvantages of the method, especially with regard to LEDs, are more often discussed [Houser et al. 2016].

5.8.4 COLOR FIDELITY INDEX R_f AND COLOR GAMUT R_g

In 2015, IES proposed a method based on the new index – the color fidelity index – R_f [ANSI/IES TM-30-15]. The method uses 99 color samples covering the color space much better than the eight (14) samples used to determine R_a. The reference source is also defined differently: for sources with a color temperature below 4500 K, the Planck source is used, while above 5500 K, the D_{55} illuminant is used. In the 4500 K–5500 K range, a linear combination of both sources is applied. A better solution – uniform color space CIECAM02-UCS – has also been proposed. The method that was developed and the applied calculation formulas provide R_f values in the range (0–100), which is also a more elegant solution. Of course, $R_f = 100$ means a perfect source, $R_f = 0$ means a source that does not render colors correctly at all in comparison to the reference source. However, there is a problem with comparing two sources of even the same value of R_f. The R_f metric tells us only about the average differences among the 99 Color Equivalent Samples from the reference, without information on how each color differs. As a result, for the same R_f, we could have a positive or negative hue shift and also increased or decreased saturation.

CIE refined the method by introducing minor adjustments and published a recommendation for R_f as an index "for scientific use" in 2017 [CIE 2017]. It is worth adding that in 2018 IES published a revised version of the description of the method based on the R_f-document ANSI/IES TM-30-18 [2018]. Currently, both CIE and IES documents are compatible. There are many publications that show that the R_f index makes it possible to assess the quality of a light source much better than the R_a index [Royer 2018; Wei et al. 2019].

In the same IES documents [ANSI/IES TM-30-15; ANSI/IES TM-30-18], another index was introduced: R_g as a measure of color gamut. This is an indicator of changes in color saturation. The same reference sources, the same 99 color samples and the same CIECAM02-UCS space are used for the determination of R_g as for the determination of R_f. The R_g value specifies the percentage of the area of gamut relation between the colors of the lights of the test source and the reference one. $R_g = 100$ means that the test source has the same area of gamut. $R_g < 100$, which means a decrease in saturation. There are also known attempts to analyze the quality of sources of light only on the basis of gamut assessment [Rea et al. 2008]; however, they have not been accepted.

On the other hand, the IES proposal [ANSI/IES TM-30-15; ANSI/IES TM-30-18] provides the opportunity to use both R_f and R_g indexes to assess the same source. An ideal source would be one for which $R_f > 90$ and $90 < R_g < 110$. However, in practice such sources are very rare – especially LEDs. If there is a low value of R_f (70–80) but a high value of R_g, this source can give attractive lighting with high saturation, e.g. in a vegetable shop, but some color distortions are to be expected. If there is a low value of R_g (70–80) but a high value of R_f, this source will illuminate with the correct colors but desaturation will occur. The objects illuminated will not look very attractive. On the other hand, if both indices have a low value (70–80), this lighting is practically only useful in places where dim lighting is expected. It is worth noting that by appropriately choosing the values of both indexes (R_f and R_g), it is possible to influence the perception of color impressions depending on the activity and tasks performed. In addition, the workers' preferences can be taken into account, which of course will be highly appreciated.

A comparison of two sources (reference and test) can be interpreted as a change in color of light that will occur after replacing the reference source by a test source. Graphic interpretation of the color change comparison for the IES TM-30-2015 method is presented in Figure 5.10.

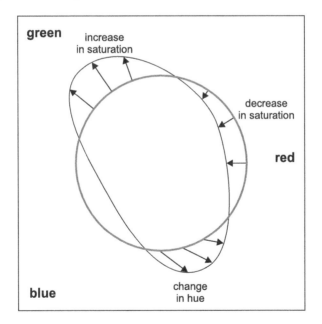

FIGURE 5.10   IES TM-30 2015 – the graphical interpretation of color change. For each color cell, the arrow indicates a change of color – from the point of the line of the reference source to the point of the line of the tested source. The arrow pointing to the center (0, 0) means a saturation decrease; in the opposite direction, an increase. However, if the arrow does not lie exactly on the straight line passing through (0, 0), then there is also a change in hue. In practice, there is usually a change in saturation and in hue.

The property of color rendition by light emitted from artificial light sources also gained significance as a quality feature after the European Union introduced environmental protection regulations (the so-called "Ecoproject"). The purpose of these regulations is to reduce electricity consumption, and hence lower environmental pollution, including a reduction of CO_2 emissions. This is regulated by Directive 2009/125/EC of the European Parliament and the Council of Europe of October 21, 2009. The rules related to lighting are described in the relevant regulations.

5.9 LIGHT SOURCES AND THEIR IMPACT ON WELL-BEING AND VISUAL PERFORMANCE

The type of light source and its properties strongly affect both well-being and visual performance. The sun – a natural source of light – has accompanied us throughout the history of human development. The first artificial light sources that had been used for many centuries were thermal ones: campfires, torches, and candles, as well as – since the 19th century – gas lights and incandescent lamps (including halogen lamps). A simple classification of light sources is shown in Figure 5.11.

We are accustomed to solar lighting, therefore the sun is a reference point for us for comparing light sources and their properties. The properties of thermal sources are similar to natural ones, although they have a lower correlated color temperature (CCT) than sunlight. People learned very quickly how to adapt thermal light sources to their needs and expectations. This state of affairs in lighting was disturbed in the 20th century by the appearance of other artificial light sources, i.e. high- and low-pressure gas discharge lamps, solid-state sources. The methods in which light

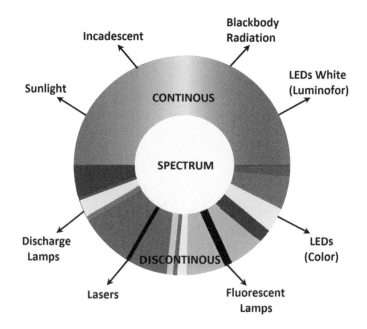

FIGURE 5.11   A simple classification of light sources.

is produced by these sources vary, which is reflected in the type of the spectra they emit and their impact on humans. The properties of such light differ from those of natural light. Despite many advantages (primarily economic ones), in many situations the use of these sources is associated with some problems [van Bommel 2019], such as flicker or the low level of the color rendering index. On the other hand, the technological development of recent years has caused one group of modern sources – solid-state ones – to change their properties in a significant way. The main advantages of these sources are durability and high luminous efficacy (lm/W). In addition, in recent years very strong economic and ecological pressure has caused solid-state sources to slowly push other sources out of the market. Incandescent lamps are practically no longer used, gas discharged lamps are slowly and systematically being withdrawn. The tendency in changing the required value of traditional light source parameters proposed by the Commission Regulations No. 245/2009 and 347/2010 [CR EC 2009; CR EC 2010] leave no doubt: in a few years solid-state sources will be the preferred artificial sources of light used by humans.

Descriptions of the properties of artificial light sources (especially the older withdrawn ones) can be found in the very rich literature of lighting technology [van Bommel 2019; Rea 2000]. It is worth considering the properties of solid-state sources in the context of lighting quality, human well-being, and visual performance.

5.9.1 CLASSIFICATION OF SOLID-STATE LIGHT SOURCES

Nowadays we can distinguish three types of solid-state light sources: LEDs (light-emitting diodes), OLEDs (organic light-emitting diodes), and PLEDs (polymer light-emitting diodes). An LED is a semiconductor optoelectronic device that emits semi-monochromatic (narrow band) radiation from a wide optical radiation spectrum range, including visible, infrared, and ultraviolet radiation. LEDs emit radiation in a narrow band which is described as a Gaussian-shape. It is characterized by two parameters: the peak wavelength corresponding to the highest emitted radiant flux, and the full width half maximum bandwidth (*FWHM*) corresponding to the difference in wavelength at which the emitted radiant flux reaches half of the maximum value. The typical value of *FWHM* is approximately 25 nm. For the production of LEDs, many different materials are used (Al, As, Ga, In, N, P, Se, Si, Zn), allowing the emission of radiation bands with different peak wavelengths [Rea 2000; van Bommel 2019].

The narrow band of LED radiation is the reason that it is perceived as light with a clearly defined color. LEDs as narrow band sources of light have very wide applications in signals and decorative lighting. However, in general lighting for workplaces or home space, we expect white light. Two methods are used to obtain such light. The first one is mixed color white light. The principle used here is mixing the light from several (three or more) LEDs of different colors, usually red (R), blue (B), and green (G) ones to create a spectral power distribution that appears white. This type of source is named RGB LED. The classic RGB is an LED strip with triple diodes (red, green, and blue on one module). The second method is phosphor-converted white light. A two-step process of generating white light is involved by using phosphors together with a short-wavelength LED. The UV LEDs, blue LEDs, or lately violet

LEDs are covered by a phosphor material. Such sources work similarly to a fluorescent lamp: the phosphor material converts short-wavelength radiation into visible radiation. By choosing the proper phosphor material for LED emissions we get the expected white light.

It is worth mentioning some advantages and disadvantages of both methods for creating white light with LEDs. Mixed-color white LEDs have good color rendering properties, high overall luminous efficacy, and flexibility for achieving any desired color property, but it is difficult to completely mix light and maintain color stability over the life and operation of the product (this concerns dimming especially). Phosphor-converted white LEDs are single compact light sources, which could be used alone or as LED matrices. By using different phosphors, we can get different CCTs of white color; nevertheless, they have limited color properties and lower overall luminous efficacy.

An OLED is an LED in which a layer of organic compound (containing carbon and hydrogen molecules) was applied. Other compounds, e.g. intrinsically conducting polymer (ICP) in PLEDs are also used. The advantage of OLEDs is the possibility of obtaining virtually any shape of the source surface. Typical applications of OLEDs are mobile phone displays and TV screens. However, due to their low efficiency, OLEDs are not used for general lighting purposes.

5.9.2   LEDs and Their Properties

Usually several areas of interest that relate to the quality of lighting are taken into account (see Chapter 4 and Table 4.1). In these areas, the different features of LEDs can be considered. In Table 5.1, the properties of LEDs as well as their basic technical parameters are described. The information provided refers to LEDs used for lighting (phosphor –converted white LEDs or RGB LEDs). Parameters (especially luminance) may differ from the given values for indicator LEDs and signaling LEDs (small low-brightness LEDs).

5.9.2.1   LED Advantages

- Mechanical and construction properties: small size, shock resistance, the ability to create virtually any spatial shape of the light source by using appropriate grids. For typical solutions, the size (diameter) of a single chip today is in the range of 1–5 mm. Their small size combined with a long life makes it possible to simplify the luminaire. There seems to be a slowly growing tendency towards compact solutions, which are integrated and disposable (without the possibility of component repairs). These advantages are very important for a lighting designer who in practice can arrange any position of sources. It is also important in decorative lighting and object illumination. The small size of LEDs in combination with their high luminance and wide possibilities of creating colors ceased to limit the creativity of illumination designers. One can only ask the question of whether this will not lead to an excess of the light effects that are proposed.
- High levels of luminance and a wide range of illuminance: nowadays 1000 lumen can be reached in constructions based on a single high-power LED.

TABLE 5.1

LED Features and Parameters Related to Quality of Lighting

Area of interest	Parameter	LEDs
Lighting intensity	Maximum luminance	5–10 Mcd/m^2 or more, depending on construction and technology
Spatial distribution of light	Illuminance uniformity of emitting surface	Possible to achieve using the appropriate module grid even for very large surfaces
	Discomfort/disability glare	Can cause glare due to very high luminance and small dimensions; new metrics for glare assessment are being developed
Color aspects	Color appearance – CCT (K)	2700–10,000
	Color rendering – R_a.	Up to 95; however R_a is not a good
	Other indexes (Section 5.8) should also be evaluated – unfortunately, manufacturers most often give only R_a values	metric for assessing LED color properties
Flicker and stroboscopic effect	–	Depending on the power supply, it is almost always present when AC power supply is used
Application and economic effect	Luminous efficacy (lm/W)	70–200 (Single LED white) 70–160 (LED module white)
	Lifetime (h)	15,000–100,000
	Warm-up time/restrike time	Instant

High luminance with a small illuminating surface makes it easy to achieve high levels of illuminance on the work surface in the workplace. However, it is worth noting that relatively high luminance (and this is a typical feature of LEDs) can create perceptual problems (see disadvantages of LEDs).

- Luminous efficacy and long lifetime: this is very important, in line with the trend to reduce electricity consumption. Cost reduction also results from the long lifetime of LEDs. It is worth noting that the reduction of costs also means no need for cumbersome (and expensive) replacement of LEDs placed in hard-to-reach places.
- Easy control option: immediate achievement of lighting parameters (warm-up and restrike time). LEDs are light sources that are easy to control, both in terms of changing the luminance level and dynamics of working (on and off). This gives additional opportunities to create dynamic lighting. Easy control of the luminance level allows wide-range dimming. Nowadays, two different methods are used: by current control or by modulation [Khan et al. 2015]. The interesting feature is that current dimming increases the efficacy of LEDs [van Bommel 2019].
- Very good color parameters: large CCT range, high color rendering index [Khan et al. 2015]. The possibility of obtaining good quality white light and

a continuous spectrum with controlled characteristics is a very important advantage. In practice, the solution with a phosphor layer is most frequently applied for this purpose. Obtaining a white color using modules of three or more individual LEDs is used very rarely. On the other hand, applications of such modules, especially when the number of individual LEDs is greater than three, provides the possibility of modeling, and in practice obtaining any shape of spectral characteristics. Of course, it is very useful in all the entertainment and decorative applications, and also in the illumination of objects. However, the most important application in this case is the ability to build sources with appropriate spectral characteristics for non-visual impact, especially taking into account the blue light component in the spectrum. It is particularly useful when we expect dynamic changes in light characteristics, e.g. for the needs of night shift work (Section 8.2). An example of a white LED with a restricted emission of blue light for night shift lighting is described in Section 8.2.3.3.

- Dimmable in a simple way: in practice, two ways of dimming LEDs are applied: either by pulse width modulation (PWM), or by constant current reduction (CCR). PWM dimming involves switching current at a high frequency from zero to the rated output current. CCR dimming means the lighting level required is proportional to the current flowing through the LED. The advantages of PWM dimming are smooth dimming capability, more precise output levels, and better consistency in color over various levels.

5.9.2.2   LED Disadvantages

- High level of luminance: generally, high luminance is an advantage of LEDs and the design of the lamp should limit the appearance of such luminance in the users' field of view. However, it is often difficult to obtain or is simply not followed by designers. In many installations, this leads to a high level of glare. In addition, such glare is very difficult to reduce.
- Thermal problems: only part of the energy consumed is converted into luminous flux in LEDs. The rest is associated with heat generation in the semiconductor structure. Junction temperature characterizes the work of a p-n junction (chip) in LEDs. Too high an operating temperature reduces the light efficiency, shortens the lifetime of LEDs, and also increases the likelihood of damage. A heat sink is used to remove heat from the chip. Thermal management of LEDs is a very important problem, especially for power LEDs [Lasance et al. 2014].
- Gradual decrease of luminous flux in time from switching on the LED: this phenomenon is mostly related to the increase of temperature in the luminaire and heat dissipation by the heat sink. Heat sinks used with LEDs are designed to absorb and disperse excess heat away from the LED and into the heat sink. Depending on the heat sink quality and luminaire construction, the increase of the LED temperature could vary and then influence the luminous flux emission. This should be especially considered during the measurements, and they should not be started just after switching on the LED.

- Flicker problem: the luminous flux is emitted immediately after switching on the LED. This is one of the basic advantages of LEDs. However, this also causes a direct transfer of flicker from the power supply. LED brightness control (dimming) can be done in impulse or analog form. In impulse form, Pulse Width Modulation (PWM) or Leading/Trailing Edge control is used. In this case, the flicker is always visible. Flicker is perceived in peripheral vision if the power supply runs below 100 Hz frequency. In the analog form, the current value is controlled (reduced). Even if this value is not constant, it does not lead to switching the LED on/off. In this case, flicker may be lower than with pulse control.
- Stroboscopic effect: using the PWM dimming stroboscopic effect could appear in fast-moving environments, when the driver frequency is low.
- The possibility of forming any spectral characteristics brings about the potential danger of photobiological hazards. Attention should be paid to the level of blue light content, which is the basis for the production of white light. This is important, because in this case, the user is not able to notice (or assess) the risk on his own and avoid it. The international standard "Photobiological safety of lamps and lamp systems" [EN 62471] defines the levels of photobiological hazard and the corresponding classes of light sources (also LEDs). Products from different manufacturers should frequently be tested to check if the requirements of the standard are met.
- Differences of production parameters, especially in terms of color. Modern technology does not allow the repeatability of LED production. Considering forward voltage, brightness, and color, similar parameters can be obtained, but they are practically never identical. To ensure repeatability, a process called binning has been developed. The manufactured LEDs are tested and classified into subgroups (bins) with the most similar parameters. Chromaticity diagrams and MacAdam ellipses are used in the binning process [van Bommel 2019].

5.10 SUMMARY

Visual performance is inherent in the quality of lighting and thus well-being. Presenting this aspect in a separate chapter is intended to emphasize its importance, especially since many of the issues involved are not directly included in the standard code.

Visual performance is a very broad term that is also associated with such human visual abilities as visual acuity, contrast perception (contrast sensitivity), distinguishing details, or the perception of light flicker. It also involves certain psychophysical laws, such as Weber's law, expressing the relationship between the physical measure of a light stimulus and the visual response in the form of a "perceptible difference", i.e. the strength of the stimulus after which the changes are visible. Taking into account how strong the influence is on visual performance of both the type of light spectrum and the way it is produced, the chapter discusses the main features of LEDs that are now commonly used for lighting purposes. The basic features of these sources as well as the advantages and disadvantages of using LEDs are presented. It can be expected that in the near future, the technological development of these sources will lead to the development of new LEDs that are of even better quality and free from some of their present disadvantages, such as flicker problems or color perception properties.

6 New Metrics for Circadian Lighting

> Providing occupants with proper circadian lighting is similar to providing them with ergonomic chairs or flat-screen computer monitors.
>
> **Mariana G. Figueiro [Bernard 2019]**

6.1 INTRODUCTION

Recent discoveries regarding the non-visual effects of light on human health, well-being, and the determination of the effectiveness of individual wavelengths in melatonin suppression and melanopsin response to light could not remain without a response from lighting engineering. How should light and lighting be assessed from the aspect of its circadian effect? Photometric parameters are used when designing or assessing lighting in the workplace and taking into account the visual effect of light. These, however, do not reflect the biological non-visual effects. Therefore, new metrics have been defined for this purpose. In the literature we can find quite a lot of different circadian metrics – and their number is growing all the time. New names (sometimes very similar to the previous ones) and formulas for their determination are proposed, and their authors try to prove the advantages of using their metric as the best one. Someone without a thorough knowledge of the field can get lost in the abundance of new ideas. In this book we present nine different approaches (and the related metrics) to determine the effect of light on the melanopsin response or the circadian system. These are both the most commonly used measures, such as equivalent melanopic lux (*EML*), circadian light (*CLA*), circadian stimulus (*CS*), and α-opic metrics, recently introduced by the CIE (Commission Internationale de l'Éclairage – English: International Commission on Illumination), but also those less frequently used, such as: universal illuminance, circadian action factor, or ones that have just been proposed for use, i.e. circadian potency, effective watts, and the melanopic-photopic ratio. What is also presented are the substantive foundations of these metrics as well as indications of their advantages, disadvantages, and interrelationships.

6.2 α-OPIC METRICS

In order to make a quantitative assessment of light exposure, the International Commission on Illumination (CIE), recommends using the light response of the five photoreceptors on the retina: S-cone, M-cone, L-cone, rod, and ipRGC [CIE 2018]. This is due to the position adopted by the CIE that a non-visual response to light cannot be described by only one function of spectral efficiency – for example, solely by the melanopsin response to light. It was thus assumed that the inputs of all photoreceptors that contribute to ipRGC-influenced responses to light should be used to characterize lighting and its effects on human health and well-being [CIE 2018]. As a result, five α-opic responses to light were obtained, which are derived from the type of opsin present in the photopigment of a given photoreceptor, denoted by the symbol α. Thus, the following response names of these photoreceptors occur: S-cone-opic (or cyanopic), M-cone-opic (or chloropic), L-cone opic (or erythropic), rhodopic, and melanopic (or ipRGC-opic). The five-α-opic action spectra for human photoreception presented in Figure 6.1 were determined on the basis of tabular data contained in the CIE system [CIE 2018]. The ipRGC sensitivity function thus presented is the effect of the consensus reached by Lucas et al. [2014]. Although the melanopic lux (melanopic irradiance) concept was originally proposed by Enezi et al. [2011], it was only later that Lucas et al. revised its melanopic function by scaling it in such a way as to ensure melanopic illuminance is equivalent to photopic illuminance for an equal-energy radiator (illuminant E). Melanopic illuminance calculated according to Enezi et al.'s [2011] method can be converted into the updated Lucas version [Lucas et al. 2014] by multiplying it by 1/5.4 [Lucas Group 2019]. An Excel-based toolbox for calculating melanopic illuminance is available on the Manchester University website under the Lucas Group tab [Lucas Group 2019].

It is worth mentioning that the curve of ipRGC response which was adopted is based on studies mainly conducted on animals. Genetically modified cells for the production of human melanopsin were exposed to short flashes of monochromatic light at different wavelengths, and the calcium response in these cells was measured. These results were used to develop the spectral melanopsin response curve.

The CIE S 026 [CIE 2018] relative spectral sensitivity of the melanopic action spectrum reaches its maximum at 490 nm, which means that this spectrum is adjusted for pre-receptoral filtering in a standard 32-year-old observer. The standard

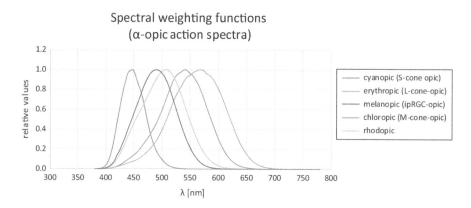

FIGURE 6.1   α-opic action spectra (based on CIE 2018).

also provides spectral correction functions for different ages, which are related to the age-dependent transmittance changes in the human eye. Please note that this action spectrum is not the same as the melatonin suppression action spectrum, which was developed based on Brainard [Brainard et al. 2001] and Thapan's [Thapan et al. 2001] studies, and reaches its maximum at 460 nm.

This standard defines α-opic-radiant flux, α-opic efficacy of luminous radiation, α-opic irradiance, α-opic radiance, α-opic efficacy of luminous radiation for daylight (D_{65}), α-opic equivalent daylight (D_{65}) luminance, α-opic equivalent daylight (D_{65}) illuminance, and α-opic daylight (D_{65}) efficacy ratio. It is worth presenting the formulas of some of the above-mentioned metrics, such as α-opic irradiance, α-opic efficacy of luminous radiation for daylight (D_{65}), α-opic daylight (D_{65}) efficacy ratio, and α-opic equivalent daylight (D_{65}) illuminance:

$$E_\alpha = \int E_\lambda(\lambda) \cdot S_\alpha(\lambda) \cdot d\lambda \tag{6.1}$$

$$E_{v,\alpha}^{D65} = \frac{E_\alpha}{K_{\alpha,v}^{D65}} = E_v \cdot \gamma_{\alpha,v}^{D65} \tag{6.2}$$

$$K_{\alpha,v}^{D65} = \frac{\phi_\alpha^{D65}}{\phi_v^{D65}} \tag{6.3}$$

$$\gamma_{\alpha,v}^{D65} = \frac{K_{\alpha,v}}{K_{\alpha,v}^{D65}} \tag{6.4}$$

where:

E_α – effective photobiological irradiance – α-opic irradiance, in W/m^2

$E_\lambda(\lambda)$ – spectral irradiance of a given test light source, in $W/(m^2 \cdot nm)$

S_α – α-opic action spectra

$K_{\alpha,v}^{D65}$ – α-opic efficacy of luminous radiation for daylight (D_{65}),

$K_{\alpha,v}$ – α-opic efficacy of luminous radiation for a test light source

E_v – photopic illuminance under a given test light source, in lx

$\gamma_{\alpha,v}^{D65}$ – α-opic daylight (D_{65}) efficacy ratio for a test light source

ϕ_α^{D65} – α-opic radiant flux of standard daylight D_{65}, in W

ϕ_v^{D65} – luminous flux of standard daylight D_{65}, in lm

$E_{v,\alpha}^{D65}$ – α-opic equivalent daylight (D_{65}) illuminance, in lx.

If we choose melanopic action spectra S_{mel} to calculate these new metrics, we get:

E_{mel} – melanopic irradiance, in W/m^2

$K_{mel,v}^{D65} = 1.3262$ mW/lm – melanopic efficacy of luminous radiation for daylight (D$_{65}$),

$K_{mel,v}$ – melanopic efficacy of luminous radiation for a test light source

$\gamma_{mel,v}^{D65}$ – melanopic daylight (D$_{65}$) efficacy ratio for a test light source

$E_{v,mel}^{D65}$ – melanopic equivalent daylight (D$_{65}$) illuminance (*MEDI*), in lx.

It is worth paying attention to a new metric, i.e. melanopic equivalent daylight (D$_{65}$) illuminance (in lx). This is illuminance produced by the radiation conforming to the standard daylight illuminant (D$_{65}$) that provides an equal melanopic irradiance E_{mel} (in W/m^2) to the test source [CIE 2018]. As daylight is the naturally occurring light stimulus to which human-centric lighting assumptions refer, this standard adopts the standard daylight illuminant D$_{65}$ as the reference illuminant to provide definitions of the α-opic equivalent daylight (D$_{65}$) illuminance. In this way the concept of melanopic equivalent daylight (D$_{65}$) illuminance $E_{v,mel}^{D65}$ (also represented in this book by the abbreviation *MEDI* when using a simpler writing nomenclature) has been applied to compare light settings to natural daylight. It should be mentioned here that although the measures adopted in this standard are based on the earlier publication of Lucas et al. [2014], the "equivalent metrics" described in this article referred to another illuminant – the equal-energy radiator (illuminant E).

The important issue is that α-opic equivalent daylight (D$_{65}$) illuminance is SI compliant. The Equivalent Daylight (D$_{65}$) Illuminance Toolbox (in Excel) version E1.051, as well as a user guide, are free and available on the relevant website [EDI 65 calculator 2019].

Using the melanopic daylight (D$_{65}$) efficacy ratio and the melanopic efficacy of luminous radiation for daylight (D$_{65}$) designated for several standard illuminants [CIE 2018] and for the model of tunable LEDs described in Section 8.2.3.2.), melanopic metrics were determined for the same photopic illuminance of 200 lx, and are presented in Table 6.1.

In summary, it could be said that α-opic irradiances are expected to be useful predictors for non-visual effects of light in human-centric lighting (HCL), especially for narrow spectral bands, mixed colors, or special whites [Schlangen 2016].

However, the measures proposed by the CIE [CIE 2018] are now more often used to describe a given light source, or to compare different light sources in terms of their effect on individual receptor responses, rather than for design. They can hardly be used in design as long as there are no requirements that specify what values of these measures should be adopted to ensure integrative lighting. Until a requirement is given for the recommended values of an individual α-opic-metric, they can serve to compare light sources and their description in research studies, but are still not very practical for designing. However,

TABLE 6.1

Melanopic Metrics for Standard Illuminants: E, A, D_{55}, and D_{65} (*) and a Model of a Localized Luminaire with Tunable LEDs at Different CCT Settings

Light source	Illuminance (photopic), E_v lx	Melanopic efficacy of luminous radiation, $K_{mel,v}$ mW/lm	Melanopic irradiance, E_{mel} mW/m²	Melanopic daylight (D_{65}) efficacy ratio $\gamma_{mel,v}^{D65}$	Melanopic equivalent daylight (D_{65}) illuminance, $E_{v,mel}^{D65}$ lx
Equal-energy illuminant E	300	1.201	360.3	0.906	271.8
Standard illuminant A	300	0.657	197.1	0.496	148.8
Standard illuminant D_{55}	300	1.199	359.7	0.904	271.2
Standard illuminant D_{65}	300	1.326	397.8	1.000	300.0
LED Model 3000 K	300	0.675	202.5	0.509	152.7
LED Model 3500 K	300	0.810	243.0	0.611	183.3
LED Model 4000 K	300	0.930	279.0	0.702	210.6
LED Model 4500 K	300	1.035	310.5	0.78	234.0
LED Model 5000 K	300	1.118	335.4	0.843	252.9
LED Model 5500 K	300	1.197	359.1	0.903	270.9
LED Model 6000 K	300	1.270	381.0	0.958	287.4

(*) Standard illuminants (E, A, D_{55}, and D_{65}) based on data included in CIE 2018.

equivalent daylight (D_{65}) illuminance has quite a good chance of being used in practice, especially if CIE and CEN (Comité Européen de Normalisation – English: European Committee for Standardization) decide to refer to it in their new normative requirements.

6.3 CIRCADIAN LIGHT AND CIRCADIAN STIMULUS

The Lighting Research Center at Rensselaer Polytechnic Institute in the US developed a non-linear model of human circadian phototransduction (the process by which the retina converts light into neural signals for the circadian system), which was first published in 2005 [Rea et al. 2005], then verified in 2010 [Rea et al. 2010], in 2012 [Rea at al. 2012], and next in 2016 [Rea et al. 2016]. The mathematical model of human circadian phototransduction is based on the neuroanatomy and neurophysiology of the retina and on psychophysical studies of nocturnal melatonin suppression using lights of different spectral power distributions. This model is based on testing people in laboratory conditions, where they have been fully dark-adapted, and then exposed to monochromatic light. It is valid both for narrow-band and polychromatic light sources and estimates the combined non-visual response of all five photoreceptors: rods, S-cones, M-cones, L-cones, and ipRGCs. It is worth mentioning that the approach for characterizing the influence of stimuli provided by light on human melatonin suppression and the human circadian system has been validated and successfully applied to assess lighting interventions in several field and laboratory studies [Rea et al. 2010; Rea et al. 2016; Figueiro 2013; Figueiro et al. 2017a; Figueiro et al. 2017b; Figueiro et al. 2019a].

The model is based on spectral irradiance distribution measured at the cornea (on a vertical plane), which is next weighted by the retinal photoreceptor action spectra (S-cone, M-cone, L-cone, ipRGC, and rod) to reflect the spectral sensitivity of the human circadian system in the acute suppression of nocturnal melatonin after 1 h of exposure. It takes into account the blue–yellow opponency, which plays an important role in melatonin suppression [Figueiro et al. 2004, Rea et al. 2004; Rea et al. 2016; Rea 2018; Spitschan et al. 2017c]. As was described in Chapter 2, this process occurs at the level of the ganglion cells. Depending on the light spectrum falling on the retina, the perception of color specific to the corresponding receptive field is being stimulated or inhibited. For example, we can perceive the color blue ("blue on") from a yellow–blue cell, when the inner receptive field (blue) is stimulated and the outer receptive field (yellow) is inhibited, and vice versa for "yellow on" (see Figure 6.2).

There are two new metrics, circadian light (*CLA*) and circadian stimulus (*CS*), that characterize the spectral and absolute sensitivities of the human circadian system. These metrics are based on fundamental knowledge of retinal physiology and the measured operating characteristics of circadian phototransduction from the response threshold to saturation. The first one, i.e. *CLA* is an irradiance weighted by the spectral sensitivity of every retinal phototransduction mechanism that stimulates the biological clock, as measured by nocturnal melatonin suppression [Rea et al. 2016].

The *CLA* model is determined by two spectral weighting functions (see Figure 6.3) [Rea et al. 2016]:

- cool light sources (i.e. b − y > 0, SPD of the light source causes output to the "blue" on) (see left picture in Figure 6.2),
- warm light sources (i.e. b − y < 0, SPD of the light source causes output to the "yellow" on (see the right picture in Figure 6.2)); b − y = 0, the output signal is "cyan".

The weighting function for cool light sources (b − y > 0) shows that for polychromatic light under certain circumstances, the total effect may be less than the effect of the monochromatic light. The circadian system does not work additively. Negative values on the

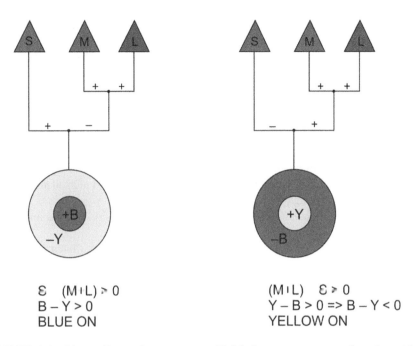

$$\mathcal{E} \ (M \mid L) > 0$$
$$B - Y > 0$$
BLUE ON

$$(M \mid L) \ \mathcal{E} > 0$$
$$Y - B > 0 => B - Y < 0$$
YELLOW ON

FIGURE 6.2 Blue–yellow color opponency (S, M, L represent cones; S – short, M – medium, and L – long).

proposed effectiveness curve for wavelengths longer than about 530 nm reflect the compensatory effect of long wave radiation on melatonin suppression [Schierz et al. 2019; Rea et al. 2016]. This statement can be confirmed by some studies on the effect of exposure to polychromatic white light with different color temperatures (of 2700 K, 3000 K, 4100 K, and 5500 K) on melatonin suppression during the night. They showed that light

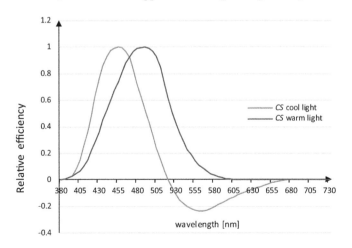

FIGURE 6.3 Spectral weighting functions for polychromatic light (for cool and warm white light sources separately), which are used in the circadian light model (based on [CS Calculator 2019]).

with a color temperature of 3000 K (warm white) and 5500 K (cool white) results in the same level of melatonin suppression (in %), while the effect of light with a CCT of 4100 K is about two times smaller than that of CCT = 3000 K, and results in comparable melatonin suppression to light as CCT 2700 K [Schierz et al. 2019]. As far as it is currently known, rods and cones participate in circadian phototransduction, and in particular the blue versus yellow color vision mechanism contributes to the spectral sensitivity of the human circadian system [Rea 2018; Rea et al. 2016]. These results are in contradiction to the predicted values of melanopic irradiance calculated using melanopic action spectra, which show higher melatonin suppression with higher CCT. To put it simply, higher color temperatures do not always mean stronger biological effects. Light sources with lower CCT can facilitate the same or even higher melatonin suppression at night than ones with a high CCT at the same lighting level [Rea et al. 2016].

The formula for circadian light *CLA* determination (6.5) consists of two main parts:

- a melanopic component (melanopic irradiance as a result of integration in the wavelength range of visible radiation of light source spectral irradiance distribution weighted by melanopsin (ipRGC) spectral sensitivity,
- an opposing cone–rod component (which is added only if the blue versus yellow (b–y) opponent mechanism is above 0); the opponent mechanism is represented by the difference between the S-cone response and the sum of L- and M-cone responses – as presented in Figure 6.2; it is worth mentioning that L and M cone response is also represented by the photopic luminous efficiency function $V(\lambda)$ [Rea et al. 2016].

The formula for circadian light *CLA* calculation is presented below:

$$
CL_A = \begin{cases} 1548 \left[\int M_c(\lambda)E(\lambda)d\lambda + \left(a_{b-y}\left(\int \frac{S(\lambda)}{mp_\lambda}E(\lambda)d\lambda - k\int \frac{V(\lambda)}{mp_\lambda}E(\lambda)d\lambda \right) - \left| a_{rod}\left(1 - e^{\frac{-\int V'(\lambda)E(\lambda)d\lambda}{RodSat}} \right) \right| \right) \right] \\ \qquad \text{if } \int \frac{S(\lambda)}{mp_\lambda}E(\lambda)d\lambda - k\int \frac{V(\lambda)}{mp_\lambda}E(\lambda)d\lambda > 0 \quad \text{condition 1: b}-\text{y} > 0 \\[12pt] 1548 \int M_c(\lambda)E(\lambda)d\lambda \\ \qquad \text{if } \int \frac{S(\lambda)}{mp_\lambda}E(\lambda)d\lambda - k\int \frac{V(\lambda)}{mp_\lambda}E(\lambda)d\lambda \le 0 \quad \text{condition 2: b}-\text{y} \le 0 \end{cases}
$$

(6.5)

where: $E(\lambda)$ – light source spectral irradiance distribution, $M_c(\lambda)$ – melanopsin (ipRGC) spectral sensitivity (corrected for crystalline lens transmittance), mp_λ – macular pigment transmittance, $S(\lambda)$– S-cone spectral sensitivity, $V(\lambda)$ – photopic luminous efficiency function, $V'(\lambda)$ – scotopic luminous efficiency function, *RodSat*: half-saturation constant for bleaching rods = 6.5 W/m², k = 0.2616, $a_{b-y} = 0.7000$, $a_{rod} = 3.3000$.

The model was normalized to the CIE standard illuminant A (CIE A 2856 K blackbody radiation). The constant 1548 sets the normalization of *CLA* to CIE illuminant A (2856 K) at 1000 lx, i.e. 1000 *CLA* = 1000 lx for illuminant A. This means that *CLA* is numerically identical to photopic illuminance, when illuminant A produces 1000 lx, but due to the non-linear character of the phototransduction model, it can differ for other spectral power distribution light sources and illuminance levels [Rea et al. 2010].

Nevertheless, *CLA* alone is not sufficient to accurately characterize the effect of light on human physiology, because it cannot be used to estimate the threshold of influence or saturation of the physiological system. For this reason, Rea et al. have developed a *CLA* transfer function to obtain a new metric (circadian stimulus – *CS*) that reflects the response of the circadian system to light by melatonin suppression. This way circadian stimulus (*CS*) is a metric determined on the basis of previously calculated *CLA*. It reflects the percentage of nocturnal melatonin suppression after a 1 h exposure, ranking the results from 0 to 0.7 [Rea et al. 2016]. The formula for *CS* is as follows:

$$CS = 0.7 - \frac{0.7}{1 + \left(\dfrac{CLA}{355.7}\right)^{1.1026}} \tag{6.6}$$

To simplify the meaning of *CLA* and *CS* metrics, it can be said that *CLA* simulates a circadian signal sent to the brain, while *CS* communicates what the brain does with this signal [Pi Lighting 2017].

The relationship between *CLA* and *CS* is often graphically represented and the resulting curve is called the absolute sensitivity of the human circadian system to light, where *CLA* is on the horizontal axis and *CS* on the vertical axis (see Figure 6.4).

However, the introduction of these metrics in practice has some limitations and complexities, such as [Wright 2018]:

- the use of *CS* requires accurate measurement at the cornea of the eye to take into account the impact of reflections from the surroundings,
- the studies on the basis of which these metrics were determined had been based only on the results of nocturnal melatonin suppression after 1 h of exposure to a specific light source in laboratory conditions, but:
  - light affects many physiological systems of the human body besides the melatonin level,
  - the duration and timing of exposure matters (i.e. it could shift or not shift the phase of circadian rhythm).

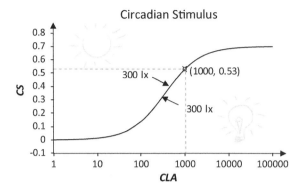

FIGURE 6.4   Non-linear relationship between *CS* and *CLA* (*CS* ≈ 0.3 at 300 lx incandescent lamp and *CS* ≈ 0.4 at 300 lx at daylight).

The value of *CLA* and *CS* can be determined using the calculator developed by the LRC for free [CS Calculator 2019]. It enables us to paste our own measurement data of the spectral power distribution of the source and calculate these metrics.

6.4   EQUIVALENT MELANOPIC LUX

Melanopic illuminance (in publications often called melanopic lux, with the abbreviation *ML*) and melanopic spectral efficiency function were proposed by researchers from the University of Manchester in an article by Enezi and colleagues in 2011 [Enezi et al. 2011]. Later, in 2014, an international group of researchers led by Lucas from the University of Manchester [Lucas et al. 2014] proposed an updated melanopic spectral efficiency function for quantifying melanopic illuminance (*ML*). The difference between the updated melanopic function and that originally proposed by Enezi et al. [2011] is that the new function has been scaled to ensure that melanopic illuminance is equivalent to photopic illuminance for a theoretical equal energy radiator [Lucas Group 2019]. Fortunately, *ML* calculated according to the Enezi et al. [2011] weighting function can be converted into the updated one [Lucas et al. 2014] by multiplying by 0.185.

WELL Building Institute [WELL 2019a] proposed equivalent melanopic lux (*EML*) as a measure of the biological effects of light on humans. It is a metric based on the melanopic sensitivity function proposed by Lucas and others in 2014 [Lucas et al. 2014] for an equal-energy radiator. The method of calculating *EML* proposed by the WELL standard [WELL 2019b] is presented below. *EML* is calculated using the melanopic ratio *R* and the expected (designed or measured) illuminance in the vertical plane at the eye cornea $E_{phot,exp}$. Melanopic ratio *R* is the ratio of melanopic irradiance (E_{mel}) and photopic irradiance, i.e. illuminance (E_{phot}) (both designated on the basis of spectral power distribution of the particular light source which is multiplied by the melanopic (ipRGC) and photopic action spectra separately). The value of *R* can be determined using the melanopic ratio calculator developed by the WELL Building Institute and available for free [WELL calculator 2019]. It is an Excel data sheet in which we can paste our own measurement data of the spectral power distribution of the source (5 nm increments).

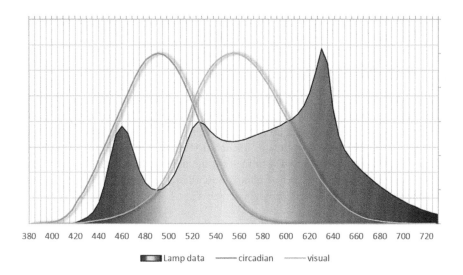

380 400 420 440 460 480 500 520 540 560 580 600 620 640 660 680 700 720

▬▬ Lamp data ──── circadian ──── visual

FIGURE 6.5 View of the graph in the WELL calculator for melanopic ratio R determination – SPD of 3500 K LED model of luminaire (see Section 8.2.3), melanopic, and photopic weighting functions (calculated $R = 0.673$).

The view of the data graphical presentation in the WELL calculator for R determination: SPD of light source, weighting functions, melanopic (circadian), and photopic (visual) is presented in Figure 6.5.

The formula for *EML* determination is as follows:

$$EML = 1.218 \cdot E_{phot,exp} \cdot R = 1.218 \cdot E_{phot,exp} \frac{E_{mel}}{E_{phot}} \qquad (6.7)$$

where 1.218 is the equal-energy radiator (CIE standard illuminant E) constant [WELL 2019b].

The example values of melanopic ratio for a few LEDs could be as follows: $R = 0.54$ for warm white 3000 K, $R = 0.77$ for neutral white 4000 K, $R = 0.93$ for cool white 5000 K, and $R = 1.1$ for daylight 6500 K. The above examples of R values for different LED color temperatures cannot be assigned to all LEDs with the same CCT [Lowry 2018]. Although the same color temperature can be determined for LEDs, the differences in spectral power distribution can significantly affect R values.

It is worth noting that *EML* is not the same measure as the melanopic irradiance E_{mel} and melanopic equivalent daylight (D_{65}) illuminance, $E_{v,mel}^{D65}$, (*MEDI*) introduced by CIE S 026 [CIE 2018]. The equivalent melanopic lux (*EML*) is based on the same ipRGC action spectrum as melanopic irradiance E_{mel} and melanopic equivalent daylight (D_{65}) illuminance $E_{v,mel}^{D65}$ (CIE S 026), but scaled differently. The unit of the measures *EML* and $E_{v,mel}^{D65}$ is the lux (lx), but they are not mathematically equal, because they use different action spectra for scaling (illuminant E for *EML* and illuminant D_{65} for $E_{v,mel}^{D65}$ (*MEDI*). As a result, mathematically: 0.91 *EML* = $E_{v,mel}^{D65}$ (*MEDI*). The

EML could also be converted to the radiometric measure – melanopic irradiance E_{mel}. Mathematically, $0.12\ EML = E_{mel}$ [in $\mu W/cm^2$].

6.5   CIRCADIAN ACTION FACTOR

Circadian action factor (a_{cv}) was developed by Gall and Bieske [Gall et al. 2004] for describing circadian efficiency of light sources with different CCTs. Firstly, they developed an averaged relative circadian action function $c(\lambda)$ with a maximum at 450 nm, based on the results of the studies by Brainard [Brainard et al. 2001] and Thapan [Thapan et al. 2001]. Then they determined the circadian action factor a_{cv} as the ratio of the integrals of the circadian and photometric quantities (weighted by $V(\lambda)$), using the same spectral distribution of the light source $X_{e\lambda}$.

$$a_{cv} = \frac{\int X_{e\lambda} \cdot c(\lambda) d\lambda}{\int X_{e\lambda} \cdot V(\lambda) d\lambda} \approx \frac{z}{y} \tag{6.8}$$

It is worth emphasizing that they present three measurement methods for determining a_{cv}, using the existing measuring devices: by spectral measurements using a spectroradiometer, by integral measurements using $c(\lambda)$ adapted detectors (they found that the blue detection channel of the digital camera of the LMK mobile photometer is similar to the circadian action function $c(\lambda)$), and, as a first approximation, to the CIE standard color-matching function [Gall et al. 2004]. The last method deserves some explanation. Comparing the action spectrum of $c(\lambda)$ with the CIE standard color-matching function $z(\lambda)$ based on standard [EN ISO 11664-1], they noticed that these are of sufficiently similar shape, which provides the opportunity for the practical application of $z(\lambda)$ in order to determine circadian quantity. Moreover, the photopic illuminance is proportional to the chromatic coordinate *y*. Therefore, as the first approximation, the a_{cv} value can be calculated from the CIE chromaticity coordinates: z and y. By converting the variables in Formula (6.8) to a_{cv}, they obtained the quotient z/y, which can be directly determined from the chromatic coordinates of a light source. This approach proved to be very helpful in radiance simulations for the assessment of circadian impact. Moreover, in practice it is easier to obtain chromatic coordinates of the source than its spectral distribution [Lowry 2018]. Besides, the a_{cv}-values correlate with CCT, which makes that metric practical. For example, $a_{cv} < 0.4$ correlates with warm white light CCT < 3000 K and $a_{cv} > 0.7$ correlates with cool white light CCT > 5300 K [Bussato et al. 2018].

This measure is useful for comparing spectral distributions of different light sources in terms of their performance [Gall et al. 2004; Pechacek et al. 2008; Bellia et al. 2011]. It also occurs as one of the metrics in the CIE technical note [CIE 2014b] related to photobiological effects.

However, the disadvantage of this measure is that it is based on a relative circadian action function $c(\lambda)$ determined for monochromatic light and its application for polychromatic light does not take into account the effect of cones and rods on the circadian response.

6.6   CIRCADIAN EFFICIENCY

Another relative measure is circadian efficiency, proposed by Bellia et al. [2011]. It differs from the circadian action factor in that circadian quantity is divided by total quantity (not weighted) for the same spectral distribution of the light source.

"The circadian efficiency in the visible field represents the potential circadian effect corresponding to a unitary radiant power emitted in the visible field" [Bellia et al. 2011]. The calculation of this metric for several sources showed that, in general, circadian efficiency increases with the rising of the color temperature CCT, with, however, some exceptions. The sources with the highest values of the action factor are those which are more efficient from the circadian point of view, under the same conditions of light emission as the luminous flux.

This metric also occurs as one of those in the CIE technical note [CIE 2014b] related to photobiological effects. However, it is not widely used in practice [Lowry 2018].

6.7   UNIVERSAL ILLUMINANCE

The concept of universal illuminance was introduced by Mark Rea, a well-known scientist in the field of lighting engineering. In his 2015 article [Rea 2015], he proposed a new weighting function for radiant energy named the universal luminous efficiency function $U(\lambda)$. As the photopic luminous efficiency function $V(\lambda)$ was developed in 1924 based on the response of only two photoreceptors (M-cone and L-cone) out of the currently known five photoreceptors characterizing the visual sensitivity of the human eye, it does not fully represent the spectral range of human visual and non-visual sensitivity to light. The new function, $U(\lambda)$, is much broader and based on spectral sensitivities of all the five photoreceptors (S-cone, M-cone, L-cone, rod, ipRGC), as presented in Figure 6.6. Like the photopic function $V(\lambda)$, it is limited in the long-wavelengths range by the L-cone spectral sensitivity curve, but in the short-wavelengths range, it is limited by the S-cone spectral efficiency curve (but not the M-cone, as in the case of $V(\lambda)$). This way $U(\lambda)$ "spans the complete range of spectral efficiencies of all known photoreceptors" [Rea 2015].

It is difficult to design lighting that supports both visual and non-visual human needs. Ensuring adequate visual performance, contrast recognition, and safety is sometimes in conflict with circadian regulation, hence the proposal of one universal luminous efficiency function, which seems to support all of the above needs [Rea 2018]. Based on the new sensitivity function, a universal luminous flux is determined (the so-called "universal lumen"), followed by a universal illuminance as the ratio of the universal luminous flux to the surface area on which the luminous flux falls. Rea proposes to use universal illuminance as the sole criterion design instead of multiple design criteria based on neuroscience algorithms. But the use of a much simpler design criterion gives "slightly" less accurate results [Rea, 2018]. Using the example of lighting requirements for classrooms and adopting the recommendation of $RVP = 0.97$ (on-axis relative visual performance – an explanation of

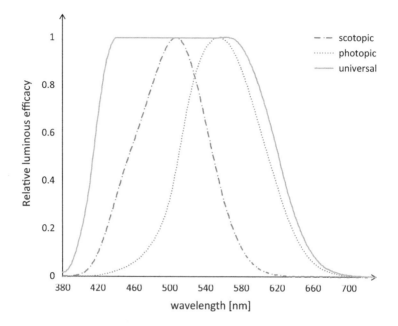

FIGURE 6.6 Relative universal luminous efficiency function versus scotopic and photopic efficiency functions. Data from Rea 2018.

RVP is provided in Chapter 5) and circadian stimulus *CS* = 0.3, calculations were carried out to determine the criteria for universal illuminance (at eye level and on the working plane). As a result of the calculations carried out for both warm and cool white light sources, the following recommendations for universal illuminance were obtained [Rea 2018]:

- 500 lx (universal) at the eye level – for proper circadian regulation,
- 300 lx (universal) on the working plane – to obtain a suitable visual performance.

In a similar way as presented in the article [Rea 2018], one can calculate universal illuminance values for other types of visual work (other *RVP* values) or other *CS* values. However, the measurement of universal illuminance can be determined either on the basis of spectroradiometric or radiometric measurements with a detector corrected to $U(\lambda)$.

Nevertheless, the one-size-fits-all concept of "universal lumen" seems to be less optimal than the multidimensional concept, which considers particular action responses of each photoreceptor [Berman et al. 2019a]. The disadvantage of the "universal lumen" resulting from its adoption of universal illuminance is that although it fulfills Abney's law of additivity and has a value of 683 lm/W at 555 nm, it is not related to any of the fundamental sensitivity functions of visual response, which means it is not a photometric quantity [Berman et al. 2019a].

6.8 RELATIVE SPECTRAL EFFECTIVENESS (*RSE*)

The other approach is presented by Amundadottir et al. [2017], who proposed a unified framework to evaluate the non-visual effect of light. They developed a new unitless metric, the relative spectral effectiveness (*RSE*) factor. This metric enables the evaluation of a spectral power distribution of light in terms of its comparative "brightness" or "energy" relationship to an equal-energy illuminant E and for any of the five photoreceptors [Amundadottir et al. 2017]. The proposed measure can be determined for both photopic vision and energy relations. The vision-related $RSE_{v,i}$ (where i could be the subscript of one of five photoreceptors) determines the relationship between the spectral weighted irradiance of a particular photoreceptor sensitivity function $S_i(\lambda)$ and the spectral weighted irradiance with off-axis CIE 10° photopic luminous efficiency function $V_{10}(\lambda)$. The example of the formula for the vision-related ipRGC relative spectral effectiveness $RSE_{v,ipRGC}$ is presented below [Amundadottir et al. 2017]:

$$RSE_{v,ipRGC} = \frac{\int_{\lambda_1}^{\lambda_2} E_{e,\lambda}(\lambda) \cdot S_{ipRGC}(\lambda) \cdot d\lambda}{\int_{\lambda_1}^{\lambda_2} E_{e,\lambda}(\lambda) \cdot V_{10}(\lambda) \cdot d\lambda} \cdot A_v \qquad (6.9)$$

where A_v is the normalization constant equal to the area under $V_{10}(\lambda)$ in the wavelengths range of λ_1 and λ_2. $E_{e,\lambda}(\lambda)$ is spectral irradiance of the test light source.

Whereas the energy-related ipRGC relative spectral effectiveness $RSE_{e,ipRGC}$ is calculated using the formula:

$$RSE_{e,ipRGC} = \frac{\int_{\lambda_1}^{\lambda_2} E_{e,\lambda}(\lambda) \cdot S_{ipRGC}(\lambda) \cdot d\lambda}{\int_{\lambda_1}^{\lambda_2} E_{e,\lambda}(\lambda) \cdot d\lambda} \cdot A_e \qquad (6.10)$$

where A_e is a normalization constant equal to $\lambda_2 - \lambda_1$, derived by integrating the area of a rectangle with a height equal to one on the interval λ_1 and λ_2 nm [Amundadottir et al. 2017].

The main advantage of this metric is that by using both vision-related and energy-related *RSE*, the absolute values can be calculated for a given irradiance or illuminance.

It is important to emphasize that *RSE* is based on the concept of equal-area normalization, assuming that until we know the role of individual photoreceptors in a non-visual system response, this approach is more consistent than normalizing action spectra to the maximum value [Amundadottir et al. 2017].

6.9 CIRCADIAN POTENCY

The new circadian metric together with state-of-the-art products (LEDs and luminaires; see Section 8.2.3) was introduced by the company CIRCADIAN ZircLight [Circadian ZircLight 2019a], based on the medical research of the scientific team led

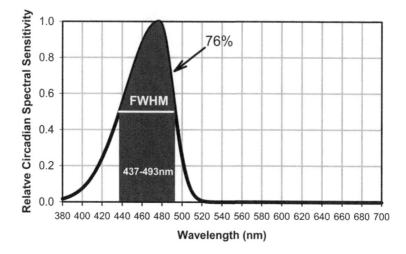

FIGURE 6.7   Relative circadian spectral sensitivity for circadian potency calculation. Taken from Moore-Ede et al. 2020, with permission.

by Martin Moore-Ede. Recognizing that the existing metrics do not have a strong correlation with human health and real working conditions, the research was carried out in a setting equivalent to typical workplace conditions, using different spectral distributions of white (polychromatic) light. A mathematical analysis of data was, therefore, carried out and a new relative circadian spectral sensitivity curve of circadian potency was determined. This new function represents a response to a steady-state stimulus [Moore-Ede et al. 2020]. The spectral range between 437 nm and 493 nm (blue light) is responsible for the steady-state response of the circadian system and 76% of the circadian potency lies in this range (shaded area under the sensitivity curve in Figure 6.7). Circadian potency spectral sensitivity curve was established for light adapted human subjects under extended exposure to white LED [Moore-Ede et al. 2020]. The shorter and longer wavelengths outside this range are responsible for a more transient response. The new metric – circadian potency is calculated "by comparing the *AUC* (calculated area under curve) of a light source in the 437–493 nm region and dividing it by the *AUC* over the entire human visible range" [Wright 2018, Moore-Ede et al., 2020].

The advantage of using this metric is the ease of its determination based on measurements with a portable spectroradiometer.

This metric has already been used in practice to evaluate the new product of the Circadian ZircLight company – "night-safe" LEDs, which emit white light with a blue light content in the 437–493 nm range of less than 2%. Thanks to this, they can be used to illuminate night shift workplaces without causing undesirable melatonin suppression.

6.10   EFFECTIVE WATTS AND MELANOPIC-PHOTOPIC RATIOS

The lack of generally accepted metrology that would meet the spectral response of ipRGCs used in both lighting and the science of vision, results in a constant search for a practical metric for melanopsin metrology. Some of the last metrics introduced at the time of writing this book are effective watts and melanopic-photopic ratios proposed by Berman and Clear [Berman et al. 2019a; Berman et al. 2019b]. Those metrics are simply and closely aligned with CIE standard unit definitions.

The first of the two proposed metrics is the effective watt (W_{eff}), which is calculated from Formula (6.11).

$$W_{eff} = \int_{380nm}^{780nm} P(\lambda) \cdot S_i(\lambda) d\lambda \qquad (6.11)$$

where $P(\lambda)$ is the spectral power of a given test light source, $S_i(\lambda)$ is the spectral weighting function of a particular photoreceptor, and each of α-opic action spectra could be used in place of i, normalized to 1.

For example, if we use the erythropic or melanopic sensitivity function, we get effective erythropic watts or effective melanopic watts. The authors argue that this metric is easier to interpret than other metrics based on lumen, because we can immediately compare the ratio of effective watts to input power in watts as a direct measure of relative efficiency. If we calculate such relations for all the individual photoreceptors, then we obtain direct measures of the efficiency of a given light source in stimulating the response of individual photoreceptors. As a result, this metric does not obtain extremely high values and is not confused with photopic or scotopic measures [Berman et al. 2019a].

Another proposed metric is the melanopic-photopic ratio (M_W/P_{lm}). It is a ratio of effective melanopic milliwatts and photopic effective watts (i.e. lumens). Therefore, the unit of M_W/P_{lm} ratio is eff. mW/lm (effective milliwatts per lumen). It is worth mentioning that this metric differs from the melanopic ratio, which is the ratio of melanopic irradiance and photopic illuminance and is usually represented by the symbol M/P, and in the WELL standard by the symbol R (see Section 6.3). As research results indicate that pupil size depends on photopic and melanopic stimuli [Tsujimura et al. 2010; Viénot et al. 2010; Schlesselman et al. 2015], the biophysical response can be described by the melanopic/photopic output ratio (level independent factor) and the photopic light level. However, it should be taken into account that results obtained by means of the above effective lux calculations are not equivalent to lux (as used by CIE DIS 026 or described in Lucas et al. [2014]). This metric is useful for estimating the relative efficiency in melanopic stimulus production from different light sources at the same lumen level. Thus, you can estimate a melanopic stimulus without having to measure with a meter properly calibrated for melanopic watts [Berman et al. 2019a]. An interesting solution presented by Berman [Berman et al. 2019b] is deriving the formula for determining M_W/P_{lm} based on S_{lm}/P_{lm} (scotopic lumens/photopic lumens). This is particularly important in a situation where the source manufacturers do not provide M_W/P_{lm} values, while they often do provide S_{lm}/P_{lm} values. Therefore, this formula allows the use of S_{lm}/P_{lm} for M_W/P_{lm} calculations without having to determine M_W (especially when they do not have a spectral power distribution of a particular light source). The formula for this calculation is presented below [Berman et al. 2019a; Berman et al. 2019b]:

$$\frac{M_W}{P_{lm}}\left[\frac{\textit{eff. }mW}{lm}\right]=\left(0.041212\cdot\frac{S_{lm}}{P_{lm}}+0.45827\right)\cdot\frac{S_{lm}}{P_{lm}}-0.07428 \qquad (6.12)$$

Some authors [Berman et al. 2019a] present a few examples of the M_W/P_{lm} application approach, such as the brightness spatial perception (which takes into account the melanopic effect, especially when the spatial brightness perception is under different spectra of light sources), pupil area control, both for foveal stimulation only and for full-field pupil response. Those examples demonstrate that the M_W/P_{lm} spectral factor could be applied in studies of lighting devoted to melanopic effects.

It is worth mentioning that the M_W/P_{lm} ratio (effective milliwatts per photopic lumen) is numerically identical to the term "melanopic efficacy of luminous radiation" defined by the CIE [2018]. A formula is also given for the calculation of melanopic lux based on M_W/P_{lm} [Berman et al. 2019b]:

Melanopic lux = (0.683/0.1621) × melanopic effective milliwatts = 4.213 × melanopic effective milliwatts = 4.213 × (M_W/P_{lm}) × photopic lux.

This new approach seems to have a chance to be successfully used for both biological and lighting research involving melanopsin stimulation by light.

6.11 SUMMARY

Most of the metrics proposed to assess the circadian effect use melatonin nocturnal secretion as a marker that is at least measurable [Lowry 2018]. However, the determination of each of these metrics requires knowledge of the spectral distribution of the source in order to use a specific efficiency curve to determine a given metric. Different ways of standardizing and scaling these metrics make it impossible to compare them with each other and only allow the comparison of different sources on the ordinal scale using a given metric. The ranking of light sources according to their potential as a circadian stimulus can, therefore, be determined only on the ordinal scale.

The circadian metrics presented in this chapter are only a selected sample from among many of those that have appeared in the literature so far. However, the metrics presented here were chosen taking into account those most often used in practice as *CS* and *EML*, and proposed by as honored a non-profit organization as the CIE, which deals with such issues as setting lighting standards, measurement methods, and terminology, representing different or interesting approaches and varied possibilities of use. There is no answer to which of these is the best, as using each of them has advantages and disadvantages.

When making use of a given metric, it is important to pay attention to the action spectrum which is used for its determination, how it is normalized and scaled, what is its real unit, whether the often-used term "lux" really refers to photopic lux, whether it fulfills Abney's law of additivity etc. Of the metrics mentioned in this chapter, only some are currently used for lighting design, as presented in Chapters 7 and 8.

7 Lighting Design Taking into Account the Non-Visual Effects of Light

Healthy light – better quality of life.

Agnieszka Wolska

7.1 LIVING IN BIOLOGICAL DARKNESS

Natural light synchronizes the internal body clock to the Earth's 24-hour light–dark rotational cycle, and the local light–dark cycles. The human body is adapted to dynamic daylight changes, which both stimulate its activity and allow relaxation. The outdoor light (daylight) on a clear, sunny day provides illuminance of approximately 10,000 lx. Indoors, in the area closest to the windows, the illuminance may be reduced by ten times, to approximately 1000 lx, but in the middle area it may be as low as 25–50 lx [NOAO 2019]. The introduction of artificial light, which often replaces natural high intensity light during the day, extends the time of wakefulness after dark and enables work at night, thus disturbing the human circadian rhythm. Both during the day at work and in the evening at home we spend over 90% of our lives indoors, mostly using electric lighting [Espiritu et al. 1994]. During the day we are exposed to light at much lower intensities than the daylight outside and during the evening at much higher intensities than the nightlight outside. The recommended illuminances for workplaces of 500 lx (and less) do not stimulate the circadian system. Since we spend a significant part of our lives at work or at home in artificial low intensity lighting, we can say that we live in "biological darkness" [de Zeeuw et al. 2019]. It can be concluded that we spend most of our time in a constant twilight that is neither day nor night [Soler 2019]. It is estimated that about 87% of the day-working population is affected to some degree by such circadian desynchronization as circadian mis-entrainment, problems with the body clock similar to jet lag, problems with sleep and alertness, and suffering from compromised mood, functioning, well-being, and health [Roenneberg et al. 2016].

As a result, the circadian rhythm phase is shifted and the timing, duration, and quality of sleep deteriorate. The chronic disruption of circadian rhythm is associated with an increased incidence of diabetes, obesity, heart diseases, cognitive and affective disorders, and also some types of hormone-dependent cancers. In order to maintain a "healthy" (not shifted) circadian rhythm, it is necessary to provide appropriate high levels of illuminance of appropriate spectral power distribution of light during the day.

Despite the proof that lighting affects both the visual and non-visual response of the human body, the design of artificial lighting at workplaces is still mainly based on the visual response. In lighting practice, it continues to be common to cater only for visual comfort, visual performance, and energy-efficiency. However, to put it simply, now it is necessary to design lighting with parameters taking into account energy-saving, visual comfort, and performance, as well as the latest findings in the sphere of the non-visual effects of light. Fortunately there are already a number of healthy lighting solutions that have been successfully used over the years. Since the early 2000s, new solutions in the sphere of lighting have been appearing which are in line with the development of knowledge in the field of non-visual effects of light on humans. Along with extending the areas of applying this lighting (not only to work areas), dynamic lighting began to include new circadian metrics. The following subsections present different types of healthy lighting solutions in historical order, ranging from the first ones related to dynamic lighting protocols to present-day solutions that incorporate new circadian metrics.

7.2   WHAT DO WE EXPECT FROM HEALTHY LIGHTING?

A number of studies devoted to the effects of lighting on the non-visual response of the human body give some guidelines for the design of "healthy" lighting. What, then, should be the characteristics of "healthy" artificial lighting that simultaneously satisfies the visual needs of a human? At first sight, it should be as similar to daylight as possible, but this works only for day shifts. Evening and night shift lighting should be designed with great care, so as not to cause negative effects on workers' health, while at the same time stimulating their alertness. A set of expected features of "healthy" artificial lighting can be summarized in Table 7.1.

7.3   DYNAMIC LIGHTING

The results of scientific work done at the beginning of the 2000s have proven that light has a significant impact on human health, well-being, performance, circadian rhythm, and mood. Striving to provide healthy artificial lighting has become a goal for researchers and lighting engineers. The idea of dynamic lighting was born in the early 2000s, just after the discovery of the novel ipRGC photoreceptors and their role in the non-visual effects of light. In the beginning, dynamic lighting was dedicated to indoor spaces with no daylight or a minimal daylight contribution. It was assumed that changing artificial lighting over specific periods of time would properly stimulate circadian rhythm and simultaneously ensure better health, the improvement of work efficiency, and lower sickness absence. The idea of dynamic lighting is still developing, adopting new patterns of lighting variability and applying new sources of light based on the developing knowledge in the area of the non-visual effects of light.

Dynamic lighting systems, the early birds of lighting engineering reflecting the latest developments in non-image-forming vision, have become professional market products available to lighting designers. Solutions of this kind are usually achieved by mixing the light output from light sources of two different (opposite) color

TABLE 7.1

Expectations from Healthy Lighting

Artificial lighting should be:	Related features of artificial lighting
Work/task dependent	Horizontal illuminance on work plane (E_h) suited to the difficulty of the visual task, providing good task performance
Work-shift dependent (day, evening, and night shift) Workday dependent Work hour dependent	Vertical illuminance at eye (E_v) and spectral power distribution of light (SPD) adaptable to circadian rhythm (intensity, timing and duration of exposure)
Geographical location and related climate dependent (tropical, temperate, continental etc.)	E_v and SPD adaptable to the local climate daylight
Weather dependent (sunny, cloudy etc.)	E_v and SPD adaptable to the weather dependent daylight
Season dependent (sunrise and sunset times, light intensity)	Timing of E_v and SPD changes adaptable to the time of sunset and sunrise
Person dependent (age, gender, chronotype, visual capabilities, health)	• E_h adequate to visual capabilities, • E_v and SPD adaptable to workers' age, gender, and/or health
Favorable for visual comfort and well-being	• Glare-free • Good color rendering • Flicker and stroboscopic effect-free • Color appearance • Modeling • Directionality of light

temperatures: cool white and warm white – on the basis of relevant optic technology, creating varying balances of cool and warm illumination. The lighting control is fully automated and in accordance with the programmed timing protocol, the color temperature and illuminance change during the day cycle.

While there are many studies done both in laboratories and in real work environments regarding the impact of different color temperatures of the light and illuminance level on the cognitive performance, alertness, mood, and well-being of workers, there are very few that used a dynamic lighting protocol.

Nowadays, dynamic lighting covers not only workplaces, but also educational facilities, nursing or retirement homes, hospitals, and our homes. The following subsections will briefly describe examples of dynamic lighting protocols in terms of their development and changes over the past two decades.

7.3.1 DAYLIGHT ORIENTED – HORIZONTAL ILLUMINANCE CONTROLLED

The aim of daylight oriented dynamic lighting is to mimic natural daylight changes in the intensity and color of light. At dawn, natural daylight is warm white and low intensity. This is followed by a gradual increase in color temperature and intensity. Light then reaches high brightness and a cool white color at noon. Then it begins to change its color and intensity again, but in the opposite direction, so that in the evening, before sunset, it reaches low intensity and warm white light again. Bluish light (like at noon) causes the biological effect of revival, stimulation for activity, while reddish light (like before sunset) has a relaxing effect. In this approach, illuminance and the blue content of the light gradually increase up to noon and then gradually decrease again towards the evening. The assumption was that this kind of dynamic lighting may be beneficial in the prevention of phase delay of circadian rhythm, which is common in the working population. Daylight-oriented protocol covers only the day shift, and especially office work. The control system of artificial lighting automatically changes light intensity (illuminance on the horizontal work plane!) and color temperature of light during the day. Examples of daylight-oriented protocols are presented in Figure 7.1. The range of illuminance changes during the day differ according to designer and user preferences: for example, it could be in the range between 500 lx and 800 lx, but also between 1000 lx and 2000 lx – stimulating the circadian system to a large degree but consuming more energy (see Figure 7.1). However, the lowest value on the horizontal plane is 500 lx, which corresponds to lighting standard requirements for office work. Nevertheless, CCT changes over time could differ. As shown in Figure 7.1, regardless of the illuminance range, the changes of CCT start from warm white light 2700 K or neutral white light 4000 K and gradually grow to cool white 6000 K or 6500 K before noon. Then they gradually decrease to warm white light. The shape of the CCT curve in the lower panel of Figure 7.1 indicates that the rise in color temperature is faster than the increase in illuminance, which is also sometimes used in practice.

Daylight-oriented protocol is one of the most commonly used methods of dynamic lighting in nursing homes for people with dementia and in senior living environments. The review article by White et al. [2013] analyzing 18 studies on dynamic lighting interventions in senior living environments concluded that this kind of time-variation lighting may mitigate the symptoms of circadian disruption in elderly people. The study of dynamic lighting intervention synchronizing light to the diurnal rhythm (direct–indirect LED system; during the day: CCT 2700–6500 K, 600–1100 lx changed gradually, during the night: CCT 1800 K) for patients with dementia in a psychiatric hospital showed that it was helpful in decreasing sleeping disturbances [van Lieshout-van Dal et al. 2019]. Another study done in a nursing home indicated that dynamic lighting in the living room (fluorescent lamps, CCT 4400 K and 389 lx during the mid-day and 1747 K and 34 lx during the night) significantly reduced agitated behavior in demented patients in a cost-effective manner [Wahnschaffe et al. 2017].

7.3.2   SUPPORTING OCCUPANTS' ACTIVITY – HORIZONTAL ILLUMINANCE CONTROLLED

A different approach to dynamic lighting aimed at artificial lighting activation to favor efficiency and well-being was developed by van den Beld and van Bommel [van den Beld 2002; van Bommel el al 2004; van Bommel 2006a; van

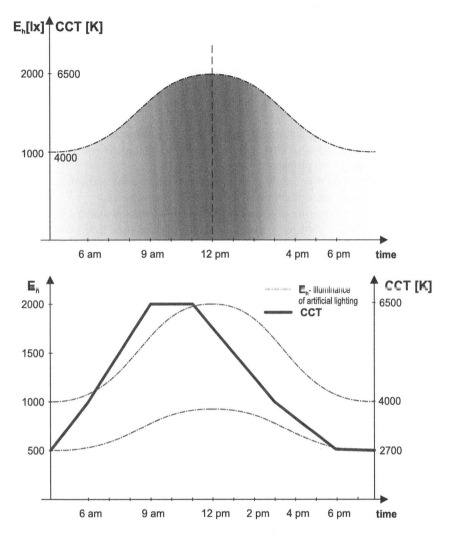

FIGURE 7.1 Daylight oriented dynamic lighting protocol: (upper panel) simultaneous change in CCT and horizontal illuminance according to the same curve; (lower panel) change in CCT and *E* according to different curves.

Bommel 2006b]. van den Beld [2002] describes his concept of dynamic lighting as follows:

- day shift lighting should start with a boosting effect, which is characterized by a high illuminance level 1000–1500 lx and cool white light, preferably more than CCT 10,000 K (about 8am). Between 10am and 1pm, lighting should be gradually dimming (a decrease of illuminance, but not to a level lower than required by the lighting standard) and changing CCT to 4000–6000 K, depending on preference, and matching with daylight contributions. After lunch, during the post-lunch dip, the level of illuminance and CCT should be gradually increased to cope with sleepiness and then again gradually increased to cope with growing fatigue at the end of the shift (especially during winter time – "winter afternoon compensation").

- evening shift lighting should start, just like for day shift, with a boosting effect of 1000 lx and cool white light and gradually reduce illuminance to the required standard level and color temperature to very low values of warm white light;
- night shift lighting should be individually controlled for the individual workplace instead of creating an overall general lighting system. Four lighting strategies were proposed:

  1st strategy – lighting resulting in no, or minimal, circadian adjustment and phase shifting – warm white light and low illuminance,

  2nd strategy – lighting resulting in partial circadian adjustment and moderate phase shifting – warm or neutral white light and immediate illuminance,

  3rd strategy – lighting resulting in maximal circadian adjustment and large phase shifting – neutral and cool white light and high illuminances,

  4th strategy – deep blue lighting to increase alertness.

Although van den Beld [2002] describes the "concept lighting algorithm" of dynamic lighting for all three work shifts, he most accurately explained the procedure of dynamic lighting for the day shift. Based on the concept developed by van den Beld, another dynamic lighting protocol (limited to day shift), was developed by van Bommel [2006a] (see Figure 7.2):

- at the time of starting work (8am) – a high level of illuminance (even almost twice as high as required by the lighting standard) and cool white color of light up to CCT 6000 K,
- in the following hours – until lunch break, i.e. about noon – the level of illuminance is gradually reduced to the value specified by the lighting standard, and the color of the light changes smoothly to a warm white light of

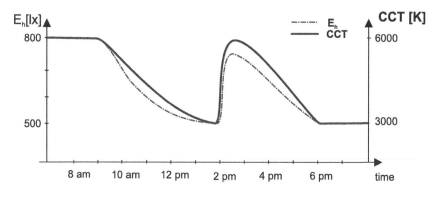

FIGURE 7.2 Lighting protocol supporting occupants' activity.

CCT 3000 K; the effect of dynamic lighting in this period of time is called the "morning boosting effect",

- at lunch time, between noon and 1pm, the lighting conditions are maintained at the same level of illuminance and CCT,
- after lunch (until 2pm) the illuminance is gradually increased to a higher level of lighting again and light color changes to cool white (up to CCT 5600 K); the aim of dynamic lighting in this period of time is to avoid the "post-lunch dip effect",
- in the following hours (until 4pm), cool white light smoothly changes to warmer again, and the level of illuminance is gradually lowered to the value specified by the lighting standard,
- before the end of work at approximately 5pm, light color is smoothly changed to cool white (called the "evening boost" [Sithravel et al. 2018] or "short boost" [van Bommel 2006a]) and just after that it is smoothly changed to warm white at 6pm. The illuminance level stays unchanged during that time.

The example of supporting occupants' activity dynamic lighting protocol is presented in Figure 7.2.

The supporting activity protocol of dynamic lighting is often referred to in scientific studies [de Kort et al. 2010; Mott et al. 2012; Sithravel et al. 2018; Smolders et al. 2009; Smolders et al. 2012; Smolders et al. 2013]. An example of research carried out in laboratory conditions was the study by Hoffman [Hoffman et al. 2008b], which compared the subjective mood and melatonin level in people exposed to dynamic (500–1800 lx, 6500 K) and static lighting (500 lx, 4000 K). The dynamic lighting protocol consisted of short peaks in illuminance in the morning and early afternoon. The study confirmed that dynamic lighting improved the subjective mood and self-reported activity, but no convincing differences in the melatonin level were found as compared to static lighting. The lack of differences in the melatonin level could probably be related to the natural low melatonin level in the morning which regardless of the type of lighting decreased to the natural daytime minimum [de Kort et al. 2010]. The other studies devoted to the influence of dynamic lighting on psychophysiological well-being were carried out in real office environments. The first large scale field study by de Kort [de Kort et al. 2010] investigated the effect of dynamic lighting (protocol according to van Bommel: 500–700 lx; 3000–4700 K) and static lighting (500 lx and 3000 K) on subjective performance, alertness, mental health, sleep quality, visual fatigue, headache, and subjective evaluation of lighting conditions. Workers experienced a certain lighting condition (dynamic or static) for three weeks. Data collection took place in the third week, after they had worked under these conditions for at least two weeks. Results showed no significant difference in the psychophysiological well-being indicators between static and dynamic lighting; however, workers felt more satisfied under dynamic lighting conditions [de Kort et al. 2010]. A potential suggested cause of such results [Sithravel et al. 2018] may be the simultaneous additional effect of daylight, as the work rooms had windows. Despite the fact that a lighting control system was used which diminished the electric lighting intensity in accordance with the changes in daylight, causing illuminance levels above 700 lx (dynamic lighting) or 500 lx (static lighting), to a greater or lesser extent daylight affected both groups of workers, whether they were exposed to dynamic or static lighting conditions.

The latest published research on dynamic lighting was carried out in windowless office rooms in the tropical climate of Malaysia [Sithravel et al. 2018]. The authors examined indicators of psychophysiological well-being during morning stimulation by dynamic lighting (morning boost) using LED lamps. Four protocols of illuminance setting were in place for two hours in the morning with constant CCT 6500 K: exposure: 500–500 lx (constant lighting), 500–250 lx (decrease), 500–750 lx (increase), 500–1000 lx (increase). The results showed that only the configurations where horizontal illuminance increased from 500–750 lx and from 500 lx to 1000 lx were supported, thus creating the morning boost. These findings indicate that dynamic lighting for the tropics requires a different protocol from that recommended by van de Beld and van Bommel. This could be related to using different protocols in the studies (especially different illuminance values and color temperatures) to reflect the differences in seasons of the year, geographical locations, and the related climate (tropical, temperate, continental etc.), which are characterized by different color temperatures of daylight during the same season (for example, cooler daylight in the tropics) and differences among the subjects (age, gender, education, and cultural and social differences of the studied groups).

To sum up the results of field research on the use of dynamic lighting as a factor improving the health and well-being of employees, it can be stated that such lighting is assessed as more satisfying by workers, but its significant impact on alertness, vitality, mental health, and performance is not unequivocal. However, the review article by Gomez and Preto [Gomez et al. 2015] concludes that the use of dynamic lighting is more favorable than that of static lighting, especially when higher color temperatures and illuminances in the morning period are applied.

The above-mentioned dynamic lighting protocols referred to the intensity of horizontal illumination, which is in some contradiction with the assumption of human-centric lighting (HCL). Since the non-visual biological effect is not dependent on the illuminance on the horizontal working plane, but on the light reaching the eye, the design and evaluation of dynamic lighting taking into account only the horizontal plane seems to be the main disadvantage. Although it is known that in the case of general lighting (luminaires installed on the ceiling), the level of illumination in the vertical plane is related to the horizontal illuminance, it is always lower than horizontal. In addition, both its level and color temperature of light (spectral composition) largely depend on the reflection properties of the room and the objects inside.

7.3.3 Diurnal Synchronization – Vertical (at the Eye) Illuminance Controlled

The studies devoted to the influence of light stimulus on the circadian system and alertness distinguish three episodes on the day timeline: the morning episode between 8am and 10am, when light exposure advances the circadian clock, the one

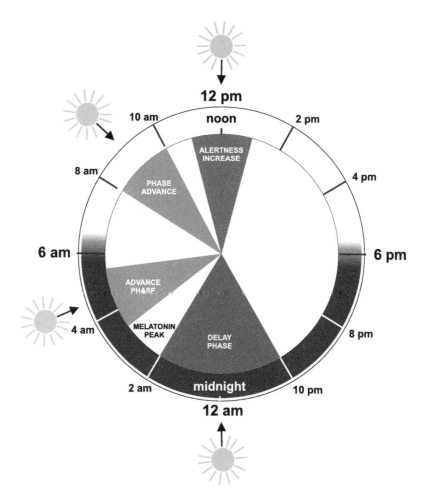

FIGURE 7.3 Timing of light exposure and non-visual effect.

around noon, when light exposure can increase alertness, and thirdly, the exposure before midnight, when light exposure delays the circadian clock (see Figure 7.3).

For human-centric lighting it is a vital condition to recommend vertical illuminances at the eye level. Such guidelines have been included in German standards DIN SPEC 5031-100 [2015], DIN SPEC 67600 [2013], and presented in: *Guide to Human Centric Lighting (HCL)* [licht.wissen 21 2018] and *Impact of Light on Human Beings* [licht.wissen 19 2014]. The following recommendations are prescribed to synchronize circadian rhythm and encourage concentration and alertness for the daylight shift (or office work):

- using illuminance at the eye between 300 and 500 lx throughout the entire work day (to synchronize circadian rhythm),
- using cool white light similar to daylight at least 5500 K for a few hours in the morning until the early afternoon,

- using warm white light with 2700–3000 K in the afternoon,
- using the daylight-oriented dynamic lighting protocol, taking into account seasonal (summer/winter) changes in daylight duration (sunrise/sunset); in the summer, the protocol runs along the natural course of daylight; in winter, the protocol extends the "sunny" day longer in the afternoon [licht. wissen 21 2018] (see Figure 7.4).
- for performance and stimulating concentration:
  - using higher illuminance levels between 1000 lx and 2000 lx on the task area, according to daylight-oriented dynamic light protocol (changes of

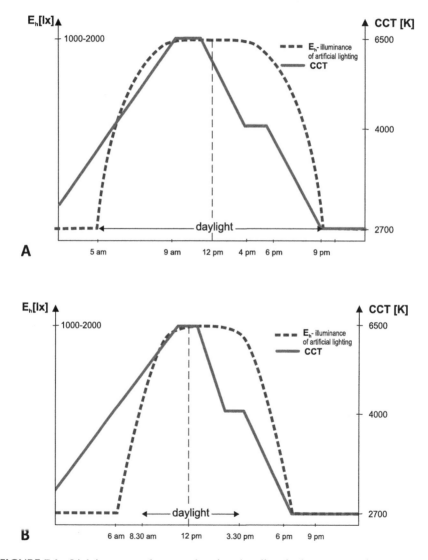

FIGURE 7.4 Lighting protocols to synchronize circadian rhythm, encouraging concentration and alertness: (A) during the summer time; (B) during the winter time.

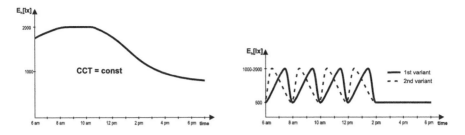

FIGURE 7.5   Lighting protocols for performance and concentration stimulation.

illuminance and color of light) achieving 250 lx illuminance at the eye
(see Figure 7.5). However, this solution is power-consuming [licht. wissen 21 2018].

- using short stimulation cycles every hour; high horizontal illuminance
  varied from 500 lx to a biologically effective 2000 lx and constant neutral white light of 4000 K, as presented in Figure 7.5. This solution is
  less power-consuming.

7.4   LIGHTING DESIGN WITH CIRCADIAN METRICS

Next to the dynamic lighting scenarios discussed in Section 7.3 which are still used
in lighting practice, while new research studies are appearing in the meantime,
new approaches to dynamic lighting design with new circadian metrics have been
developed.

Two "leading" metrics have been used in lighting design so far: one based on the
circadian stimulus (*CS*) (see Section 6.2) and the other one based on the melanopic
lux (*ML*). Both are used to determine the effectiveness of various light sources in
inhibiting melatonin secretion. However, the different phototransduction models that
were adopted, which use different spectral sensitivity weighting functions of melatonin suppression may in practice give conflicting guidance on human-centric lighting.
There are two distinct scientific groups behind the leading metrics: circadian stimulus
(*CS*) was developed at the Lighting Research Center at Rensselaer Polytechnic [Rea
et al. 2005; Rea el al. 2010; Rea et al. 2012; Rea et al. 2016], while melanopic lux
(*ML*) was developed at Manchester University [Enezi et al. 2011; Lucas et al. 2014].

Currently there is no agreement in the scientific and designer communities
regarding the use of only one of these measures. There are supporters of both. It can,
therefore, be said that the two measures "compete" with each other. A survey conducted in 2017 by *Metropolis Magazine* [Clark et al. 2017] showed that when asked:
"Which circadian metric do you use in practice", 62% of the respondents answered
that they used *CS*, whereas the remaining 38% used *ML*. However, given that the
Draft International Standard CIE DIS 026: 2018 [CIE 2018] concerning metrology
of optical radiation for ipRGC-influenced responses to light was released in 2018,
and the adopted melanopic action spectra were developed by the Lucas Group, it can
be presumed that the proportions of the users of both measures could soon change.

7.4.1 Circadian Stimulus (*CLA* Model) in Lighting Design

The light stimulus that is transformed into neural signals through circadian photo-transduction is called circadian-effective light [LRC 2019a]. Lighting characteristics affecting the non-visual response (circadian system) are different to those affecting visual response (visibility) [LHA 2019]. As its name implies, circadian stimulus (*CS*) quantifies the circadian system response after 1 h of light exposure. Therefore, the choice of its value, which corresponds to the melatonin suppression, should be properly selected for the time of day, so as not to cause undesirable rhythm phase-shifts on the one hand and to stimulate alertness and efficiency on the other. Considering the above, lighting with a circadian stimulus is characterized by the variability of illuminance, or of both illuminance and light color (spectral composition) in such a way as to obtain the desired *CS* values in the correct timing and of correct duration. Researchers from the LRC have developed basic requirements and protocol for dynamic lighting with a circadian stimulus. For a period of normal daily activity of a person working in the daytime (from coming to work to the evening before going to bed), it is recommended changing both illuminance and color of light, applying the following general schedule [Figueiro et al. 2017b; Pedler et al. 2017]:

- stimulation of the circadian system in the morning up to 11am (for maintaining proper entrainment with the local time on Earth), using higher circadian stimulus $CS \geq 0.3$ (by bright cool white light or blue light; however, it is also possible to use bright warm light at 3000 K),
- gradually lowering the circadian stimulus (by using gradually warmer and less bright light, or only making it less bright) up to low $CS < 0.1$ in the evening at about 7pm (dim warm white light).

As mentioned above, recommended changes in *CS* values during the day can be obtained in two ways:

- by changing the color temperature of light and the illuminance level (especially recommended for higher *CS*). This solution demands a color tunable lighting system, which requires higher investment costs for the purchase of a suitable lighting system, but during operation it involves relatively low energy costs;
- by changing only the level of illuminance without changing the color temperature of light. This solution demands only a dimming system, which requires relatively low investment costs, but during operation relatively higher energy costs are incurred, due to the need to ensure higher levels of illuminance in order to obtain high *CS* values.

Another important aspect in lighting design is minimizing the use of energy. If any of the above options should be chosen, the energy costs necessary to obtain high *CS* values should be considered. Therefore, it is essential to choose a luminaire with an appropriate luminous intensity distribution to obtain the required *CS* values and illuminance on the work plane. It is recommended applying distribution of light with

a vertical (E_v) to horizontal (E_h) illuminance ratio of at least 0.7 [Bernard et al. 2019]
However, the latest study by Jarboe et al. [2019] showed that an even lower ratio of
0.65 permits achieving $CS \geq 0.3$ while employing lower illuminance. As a result, a
more energy-efficient solution is achieved. Comparative analysis of the effectiveness of the three main types of light distribution (direct, indirect, direct–indirect)
showed that direct–indirect distribution has the best E_h to E_v ratio [Figueiro et al.
2016b]; however, direct distribution could also provide an adequate ratio [Jarboe et
al. 2019]. Considering that in the group of direct–indirect luminaires, there are both
different proportions of direct–indirect components (in assigned ranges) and different shapes of luminous intensity curves, it was pointed out that it is important to
choose the appropriate luminaire from this group. It is recommended ensuring a possibly wide-beam in the lower half-space (direct luminous flux not bigger than 50%)
and a significant luminous flux emitted in the upper half space (indirect luminous
flux of at least 50%) [Figueiro et al. 2016b]. The situation is similar in the group of
direct luminaires. A more detailed analysis of luminaires of different light distribution showed that the best E_v/E_h ratio was obtained using a 2 × 4 troffer luminaire of
direct wide distribution in the lower half space. Using this luminaire with warm white
light 3000 K, it is possible to achieve $CS = 0.30$, but with cool white light 6500 K,
it is $CS = 0.40$ (both for 300 lx on task area) [Jarboe at al. 2019]. In turn, Dai et al.
showed in their lighting simulations that the use of indirect luminaires on a highly
reflecting ceiling is beneficial both because of providing high indirect corneal illuminance derived from reflections, and because of a much more uniform distribution
of corneal illuminance than when using direct luminaires [Dai et al. 2018].

7.4.1.1   Examples of Dynamic Lighting Schedules

There are several proposals for dynamic circadian lighting schedules available in
LRC publications. Unlike the dynamic lighting protocols developed by van den Beld
and van Bommel [van den Beld 2002; van Bommel 2006], no bright cool white
light (as a boosting effect) is used either during the post-lunch dip or in the evening
before the end of work. Circadian stimulus changes during the day, which gives a
kind of dynamism to this lighting, providing a prompt to call it dynamic lighting
with a circadian stimulus solution. It should be emphasized here that a dynamic
lighting schedule, which has to adjust the illuminance level (luminous flux output)
and light color delivering the circadian stimulus throughout the workday, requires
programmed tunable-white LED lighting systems. An example of a hypothetical
dynamic lighting schedule (with illuminance and CCT changes in specific time
intervals) for the time range between 6am and 6pm, developed with emphasis on
occupational health and well-being is presented on Figure 7.6 (based on Figueiro et
al. 2016b; Pedler et al. 2017]). The Authors emphasize that the schedule they present
should not be interpreted as a general lighting prescription. Another example of CS
changes (by tuning the CCT and illuminance level adjustment) can be found in other
articles [Figueiro et al. 2016b; Pedler et al. 2017]).

Figure 7.7 presents yet another example of a hypothetical dynamic lighting schedule with illuminance changes only and constant warm 3000 K light, in the hourly
range between 6am and 6pm. The horizontal illuminance is shown on a percentage
scale from 100% at high CS to 19% at low CS. The choice of warm light is associated

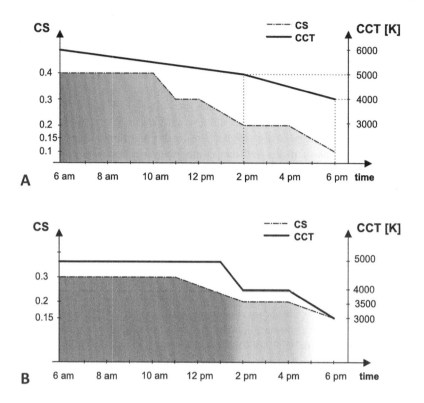

FIGURE 7.6 Dynamic lighting schedule with *CS* and CCT changes: (A) with a mild gradual decrease in CCT over time to 4000 K and gradual decrease in *CS* in the range from 0.4 to 0.1; (B) with a gradual decrease in CCT over time to 3000 K and *CS* in the range from 0.3 to 0.15.

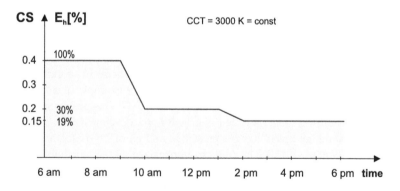

FIGURE 7.7 An example of dynamic lighting protocol with *CS* metric – changes of *CS* by changes of light output (illuminance) in % only (CCT – 3000 K – const).

with the need to use higher levels of illumination to obtain high *CS* values in the morning, compared to the use of cool white light, e.g. 6000 K. As a result, this solution is more expensive due to higher energy consumption. On the other hand, when using warm light, it is easier to obtain low *CS* values in the afternoon by simply reducing the illuminance level. However, choosing this option in interiors with good access to daylight, we will get a combination of warm and cool white light, which creates an inappropriate lighting mood. It is, therefore, better to use light sources with at least a white neutral color (4000 K).

7.4.1.2 Desktop Lighting for Delivering Circadian Stimulus

Yet another solution is making the choice to equip work stations with programmable tunable desktop luminaires specifically dedicated to providing the appropriate circadian stimulus at specific times of the day. This kind of luminaire could be placed in front of the worker, so that the light is directed towards the eyes. This solution seems to be a priority application wherever it is not possible to modernize or replace the existing general overhead lighting.

The examples of scientific-based programmable tunable LED-based desktop luminaires were developed by the Lighting Research Center (LRC). What makes these luminaires different from others available on the market are the following:

- they were developed on the basis of scientific research,
- they were applied in real workplaces, and wide-ranging studies were carried out,
- the study results confirmed that it is possible to get adequate *CS* at the right time of day and improve the occupants' alertness and sleep onset,
- the methodology and results of studies have been published in scientific articles, presented at conferences, and spread widely in many popular-science materials available on the LRC website.

The first of the desktop luminaires that were designed has two variants: 1) equipped with blue LED strips (470 nm) or 2) equipped with cool white LED strips (6000 K), both of the same dimensions and covered by the same diffuser. Both variants of luminaire provide *CS* > 0.3. They could be placed either below the participants' computer displays or above the display, according to user preference. Both variants of the luminaire were tested in field studies in four independent office sites in the U.S. (two federal government office sites and two U.S. embassies). The results confirmed that this type of lighting delivers *CS* > 0.3 in the morning and it improves the acute alertness and vitality of the workers [Figueiro et al. 2019a].

The second of the desktop luminaires developed by LRC (designed to deliver saturated blue light or white light, or saturated red light) is presented in Section 8.2.3 because it seems to be more convenient for readers to relate it with shift work lighting.

7.4.2 EQUIVALENT MELANOPIC LUX IN LIGHTING DESIGN

The WELL Building Institute standard proposed the following requirements for *EML* in work spaces, of which at least one should be met [WELL 2019b]):

- At 75% or more of the workstations, $EML \geq 200$, measured on the vertical plane facing forward, 1.2 m above the floor (to simulate the view of the occupant). This value of EML may incorporate daylight, and is present for at least 4 h between 9am and 1pm, for every day of the year; this value may be lowered after 8pm at night [WELL optimization 2019]
- For all the workstations, electric lighting (which may include task lighting) provides maintained illuminance on the vertical plane facing forward (to simulate the view of the occupant) of $EML \geq 150$.

The new optimization concept for circadian lighting design has been posted on another WELL website devoted to the principles of Version 2 of the WELL standard, i.e. WELL v2.Q3 [WELL optimization, 2019]. This concept relates to alternative use when designing circadian stimulus CS instead of EML and reads as follows: "projects may use a circadian stimulus (CS) threshold of 0.3 from electric light, since a CS of 0.3 corresponds to between 150 and 315 EML for most commonly used light sources". Although this concept was approved on February 1, 2019, it has not yet been mentioned and/or included in the scope of the WELL v2.Q3 standard. Designers who use only the recommendations presented in the content of the standard do not know this alternative recommendation at all, and therefore it is not used in practice. In addition, it should be noted that the recommendation of alternative CS use applies only to the design of electric lighting. Thus, the project which is to include daylight still refers only to the EML metric. However, there are no lighting requirements that could be used for shift work. Therefore, when designing circadian lighting for shift work in industry, it is not possible to use this standard for night shifts.

Nevertheless, although the EML metric has been adopted in the above architectural standard [WELL 2019a], it has not been sanctioned by any independent standard organization (like CEN, CIE, or ISO) or research study [Jarboe et al. 2019]. One advantage of this standard is that it boldly tackles the difficult subject of human-centric lighting. It can, however, be accused of lack of documented research confirming the proposed requirements for EML, nevertheless, currently the proposed values are also associated with the requirements for CS. Over the past two years, the requirements of this standard for circadian lighting have changed so as to take into account the development of knowledge in this area, and, e.g. the requirements have been reduced from $EML \geq 250$ (equivalent to 226 lx for daylight by CCT 6500 K) to $EML \geq 200$, which is equivalent to $CS = 0.3$ for daylight by CCT 6500 K at 180 lx [Konis 2017]. It is again worth mentioning that using this metric only considers the ipRGC response, and not the human circadian system.

7.4.3 Comparison of CS and EML

As for the use of new metrics for human-centric lighting design (as presented in Section 7.4), 2018 survey results showed that the CS metric is more often used than EML. Nevertheless, in fact there are still many more designers who do not use either of these metrics. The two concurrent metrics have their supporters and opponents. The opponents of ML/EML say that the melanopsin action spectrum (actually based on photopigment response, which generates neural response after absorbing light) is never the same as the spectral sensitivity of a neural channel response measured with

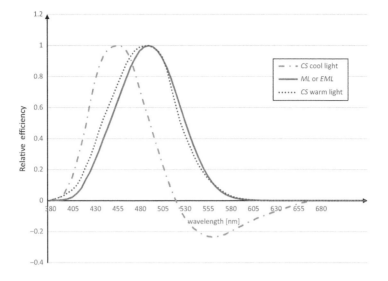

FIGURE 7.8 Weighting functions for *CS* and *EML* (*ML*) determination (based on [CS Calculator 2019; WELL Calculator 2019]).

human participation through registrations using electrophysiological methods (EEG, EOG, EKG etc.) and behavioral methods. Others argue that the use of *CS* only makes sense when used to design artificial lighting. Besides, it should be noted that some of the researchers argue that both models are not sufficiently complete for adequately determining circadian effectiveness. The main argument is that none of the models come from real life and work conditions, but only from laboratory tests conducted either on animals (*EML*), or on people (*CS*). Moreover, the existing measures do not have a strong correlation with human health [Wright, 2018]. Weighting functions for *CS* and *EML* determination are presented in Figure 7.8.

Table 7.2 presents a comparison of the main features of the *CLA/CS* model and the *ML/EML* model that can be helpful when choosing a specific model for practical use.

New proposals for circadian metrics are still appearing (see Chapter 6) and also have their supporters and opponents. It is difficult to predict when this "competition" will end and finally one metric is adopted for human-centric (integrative) lighting that will satisfy both the scientific community and lighting designers.

As a result, no circadian metric has been implemented in international lighting standards so far. Such an outcome seems to be caused by a conservative attitude, i.e. avoiding decisions due to the possibility of making some possible mistake or creating a problem that could cause potential harmful effects. Therefore, the first principle is: do no harm. Rea and Figueiro stated, "However, avoiding the problem also avoids any errors of omission (do not be negligent)" [Rea et al. 2016]. On the other hand, it can also be stated that the current knowledge is sufficient and verified in many studies and practical applications, so it is already possible to formulate requirements, and "avoiding the problem is much more likely to lead to errors of omission than errors of commission" [Rea et al. 2016]. It follows that in the standard lighting requirements for HCL design we are stuck somewhere between "do no harm" and "do not neglect".

TABLE 7.2

Comparison of the Main Features of *CLA/CS* and *ML/EML* Models

Feature	*CLA/CS* model	*ML/EML* model
Action spectrum	• Two spectra: one for cool and one for warm light sources (see Figure 7.8) • Based on melatonin suppression in humans • Effect of interaction of all retinal photoreceptors included	• Based on melanopsin response to light in studies on animals (see Figure 7.8) • No interaction with other retinal photoreceptors included
Metric interpretation	Quantifying melatonin suppression level caused by light source to determine adequate illuminance level	Measuring biological effect of light source on humans at particular illuminance level
Linking to circadian output: melatonin suppression and phase shift	Yes	No
Linearity with illuminance level	No. Non-linear dependence of *CS* on *CLA* illuminance level	Yes. Linear dependence of *EML* on illuminance level
Scientific basis for lighting recommendation	Yes. Based on studies with humans and nocturnal melatonin suppression after a 1 h exposure to light of different spectra	• No information • *EML* recommendations probably based on *CS* recommendations
Recommendation for day shift lighting	*CS* ≥ 0.3 for at least 1 h recommended before 11am	200 *EML* for at least 4 h between 9am and 1pm
Comparison of *CS* and *EML* at *CS* = 0.3	6500 K daylight SPD at ~180 lx equivalent to *CS* = 0.3 [Konis 2017]	200 *EML* – equivalent to 6500 K daylight at ~182 lx [Dai et al. 2018; Brennan et al. 2018]
The same metric value for the same CCT of different light sources	No	No
Application in field studies	Yes.	Yes
The need for measurements at the eyes	Yes [Wright 2018]	Yes [WELL 2019]
Application for artificial and daylight design	Yes/no Adaptable for daylight [Hartman et al. 2016] Only for artificial [Brannan et al. 2018]	Yes Both for artificial lighting and daylight [WELL 2019a]
Recommendation for dynamic lighting	Yes	No
Recommendation for night shift work lighting	Yes	No

(*Continued*)

TABLE 7.2 (CONTINUED)

Comparison of the Main Features of *CLA/CS* and *ML/EML* Models

Feature	*CLA/CS* model	*ML/EML* model
Additivity	No [Rea et al. 2012; Dai et al. 2018]	Yes
Smart control	Required	Required
Model development	Lighting Research Center at Rensselaer Polytechnic, US	"Lucas Group" at Manchester University, UK
Main promotional organization	The Light and Health Alliance	WELL Building Institute

7.5 SUMMARY

Designing lighting that includes the non-visual effects of light is already quite widely reflected in both scientific articles and CIE publications as well as in design practice promoted by many lighting companies, entities issuing certificates for so-called "healthy buildings" (such as the WELL Building Institute), professional journals (such as licht.wissen), or architects and lighting practitioners. The principles and practical examples of designing healthy lighting presented here in relation to single-shift operations between 6am and 6pm constitute a representative slice of the existing and described solutions. The use of any of the solutions presented may contribute to improving the well-being and health of workers. It is difficult to say which would be the best one, because the choice of lighting also depends on economic considerations (installation costs, operating costs etc.) and the features of work activities, as well as the characteristics and expectations of the users (workers).

Even the use of the simplest version of dynamic lighting, which appropriately changes only the level of illuminance over time, without changing the color of the light, is a good and relatively low-cost solution. It should be remembered that the selection of a luminaire with optics ensuring a vertical to horizontal illuminance ratio of at least 0.65 is important in order to achieve adequate circadian metrics and an energy-efficient solution [Jarboe et al. 2019]. In addition, when choosing luminaires, it is worth considering those that have a significant share of the indirect component and at the same time ensure a high ceiling reflectance, so that the cornea of the eye has a higher illuminance and uniformity. Nevertheless, if possible, more sophisticated solutions should be encouraged, especially where there is limited access to daylight.

More guidance on the design of healthy lighting is provided in Chapter 9.

8 The Biology of Shift Work and the Role of Lighting in the Workplace

Improvements in productivity, a decrease in accidents, an increased level of mental performance, improvements in sleep quality, and an increase in morale among night shift workers have been attributed to better lighting conditions.

Marcel Harmon [Harmon, 2019]

8.1 SHIFT WORK AND HEALTH

Since we live in a globalizing world, the impact of shift work has been significantly increasing in recent years. The comparison of shift work in 2000 and 2014 showed an increase of 3–14% in subjects reporting working a day shift, evening shift, or night shift [Cheng et al. 2018; Cheng et al. 2019]. According to one definition of shift work, the last portion of the shift occurs between 7pm and 6am [Caruso 2014; Wickwire et al. 2017].

Shift work implies the misalignment of the biological time in relation to real time. Employees are required to work at times when sleep typically occurs and to sleep during the daytime, which is usually the time of activity. In humans the sleep–wake system is governed by an internal biological clock. As previously described in Chapter 3, Section 3.2, the molecular circadian clock is present not only in the cells of the central circadian clock in the suprachiasmatic nucleus (SCN) but also in the peripheral cells of all tissues (Figure 8.1).

The peripheral cell-specific molecular circadian clocks show the 24 h rhythmic expression of the clock-genes. Circadian dysregulation alters the normal synchrony of all the clock-components of all living systems, including human beings, with implications for health/well-being and disease. The problems that can occur relate to physical and mental health, safety, social life, and work performance [Saksvik et al. 2011].

8.1.1 SHIFT WORK TOLERANCE

Nevertheless, it is important to consider individual differences in tolerance to shift work: some people tolerate shift work well, whereas others develop serious health problems in response to such exposure. What does the term "shift work tolerance" mean? This problem was first tackled by Andlauer et al. [1979], who defined it as the ability to adapt to shift work without adverse consequences, probably associated

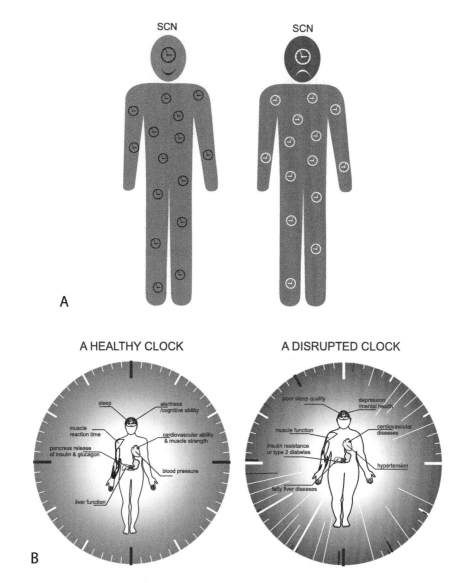

FIGURE 8.1   Synchronization (left) and desynchronization (right) (A) of central (SCN) and peripheral clocks; (B) biological effects of circadian desynchronization.

with behavioral and biological/genetic predispositions. Reinberg and Ashkenazi [2008] determined the following symptoms of shift work intolerance:

- persisting sleep alterations/disorders and fatigue, responsible for regular use of sleep medication,
- changes in behavior (increased aggression and sensitivity),
- digestive problems.

Other authors also point out the occurrence of cardiovascular problems [Klawe et al. 2005].

Saksvik et al. [2011] reviewed the published literature regarding the relationship between individual differences and different measures of shift work tolerance. They concluded that young age, male gender, low scores on morningness, high scores on flexibility, low scores on languidity, low scores on neuroticism, high scores on extraversion and internal locus of control, and some genetic predispositions are related to higher shift work tolerance. Morningness/eveningness (the biological chronotype) seems to be a genetically linked trait. This term refers to the period during the day when subjects are most awake and active. "Morning larks" (morning chronotype) are individuals who are most functional in the early morning hours. "Night owls" (evening chronotype) are more active during the late afternoon or evening hours. A shift of circadian deep temperature curves was also observed between both chronotypes, with the minimum occurring earlier in morningness. Morning and evening preferences are linked to the period of the circadian clock "and a length polymorphism of the *PER3* clock gene" regulating sleep and wakefulness [Taniyama et al. 2015; Wickwire et al. 2017]. The last study from the Hattamaru team [Hattamaru et al. 2019], who examined clock gene expression in Japanese shift-working men, suggests that night shift work affects the rhythms and levels of circadian *PER3* and *Nrld2* expression dependent on the shift schedule or type of shift.

Circadian preference – biological chronotype – has been hypothesized to modulate tolerance for shift work, where the morning type exhibits reduced tolerance [Taniyama et al. 2015]. Physiological aging and neurodegeneration in the SCN (resulting in the shortening of the circadian phase and in the preference for morningness) seem to be involved in age-related reduction of the tolerance for shift work [Wickwire et al. 2017]. It follows that young people (under 30 years old) and the evening chronotype are better biologically adapted to shift work, especially to night shifts.

8.1.2 Shift Work Disorder

Sleep disturbance and sleepiness are the basic criteria for shift work disorder. As determined by the International Classification of Sleep Disorders ICSD-3 [AASM 2014], "shift work disorder is characterized by insomnia (when sleep is allowed) and/or excessive sleepiness during wakefulness, typically accompanied by a reduction of total sleep time" [Cheng et al. 2019]. The symptoms should be present and associated with the shift work schedule for at least three months, and cause clinically significant impairment in mental, physical, social, occupational, education, or other important areas of functioning. Borbely et al. [2016] proposed two process models of sleep regulation: the tendency to sleep is regulated by the interaction between "sleep pressure" (increasing with each hour of wakefulness and decreasing with sleep) and the "circadian alerting signal" typical for wakefulness, regulated in the SCN. The light/dark cycle regulates circadian processes. Light entering the eyes results in the suppression of melatonin secretion, while the onset of melatonin secretion under conditions of low light is a marker of the physiological night (as described in the physiological part of this book). There are also important health consequences beyond sleep disruption and sleepiness. These are:

shift work-evoked uncoupling sleep pressure and the circadian alerting signal. The following medical complications can be listed: increased risk for cardiovascular disease, cerebrovascular events and stroke, poor sexual health, multiple forms of cancer [Wickwire et al. 2017]. Hunger, food preference, and metabolism have a strong circadian component. Unhealthy eating, bad meal timing, and circadian variations in energy metabolism might be involved in the progress of obesity and increase the risk for being overweight related to the duration of shift work [Cheng and Drake 2019]. The next consequence of mistimed eating seems to be reduced glucose tolerance and a moderate increase in the risk for diabetes among shift workers [Knutsson et al. 2014]. On the other hand, Itani et al. [2017] showed that high-risk lifestyle factors such as short sleep duration, shift work, and actual days taken off work are predictive risk factors for new-onset metabolic syndrome (the cohort study was carried out on a group of 40,000 male workers). Other medical consequences of shift work include an increase in risk for the development of specific forms of cancer [Yuan et al. 2018]. In the characteristics of shift work disorder (see above), a decrease in the cognitive domain was listed. The exposure to shift work over less than five years seems to be reversible. A clinical study on night shift nurses indicates that depression is a problem in shift workers. One possible explanation of the affective functioning is interaction between melatonin activity and monoaminergic activity (involved in mood regulation): melatonin activity may have downstream effects on mood [Cheng et al. 2019]. Mood disturbances are exacerbated by impaired psychosocial functioning of shift workers; social isolation and reduced social participation resulting in a higher risk for somatic and psychiatric disorders. Night shift work increases cold pain perception [Pieh et al. 2018]. The same cold pain stimulus is rated 28% more distressing after night shift work and normalizes after a recovery night. Increase in cold pain perception after night shift work appears to be more related to changes in mood in comparison to changes in sleepiness [Pieh et al. 2018]. Kim et al. [2017] used magnetic resonance image acquisition and voxel-based morphometry to show a regional grey matter volume reduction of pontomesencephalic tegmentum – in the region corresponding to the nuclei that consist of major aggregation of cholinergic neurons, considered a part of reticular activating controlling cortical activity system across the whole sleep–wake cycle [Saper et al. 2010; Kim et al. 2017]. The authors speculate that their findings "may be related to chronic disruption of circadian rhythm or to decreased exposure to the bright light in shift workers".

All the medical, psychological, and social consequences of shift work and/or night shift work that have been described force us to find specific clinical recommendations for patients who are shift workers: elimination of shift work schedule, regular assessment (mood, cancer, sleep-related comorbidity, and other forms of comorbidity), development of a personalized treatment plan (to reduce sleepiness and improve sleep), and family and social support [Wickwire et al. 2017].

Evidence suggests the positive role of light therapy, especially bright light therapy or using other *zeitgebers* as potentially modifiable regulators of circadian rhythmicity, for example exogenous melatonin administration; however, strong control of exposure to light must be observed [Wickwire et al. 2017]. Light has great potential to positively affect health and performance, but it also presents a possible health threat. Short wavelength light can increase the risk for age-related macular degeneration, and blue lighting at night is another potential health risk [Westland et al. 2017]. Thus, workplace lighting during shift work has two important tasks to fulfill: on the one hand to support alertness, cognition, and performance during work, and on the other to improve mood and reduce negative health consequences in individuals who work day, evening, and night shifts. It will be important to consider the biological effects of lighting spectrums in home and workplace design.

8.2 SHIFT WORK LIGHTING

8.2.1 DEALING WITH CONFLICTING VISUAL AND NON-VISUAL NEEDS

Shift work, particularly that including night shifts, is the most widely studied condition, as it may interfere with human homeostasis and well-being at several levels [Costa 2010]. Workers must face a significant shift in the circadian phase and the resulting consequences, both for their health and due to problems related to family and social life. The stress arising influences workers' general well-being as well as affecting both their physical and mental health [van den Beld 2002].

Light of suitable spectral composition and intensity used at the proper time may adjust the timing and amplitude of circadian rhythms [Arendt 2010]. The main parameters of circadian rhythm are presented in Figure 8.2.

In general, light treatment during the "biological night" [Arendt 2010] might cause:

- phase delay (see Figure 8.3) – using alerting cool white light during the first half of the night, before the melatonin peak,
- phase advance (see Figure 8.3) – using alerting cool white light during the second half of night, after the melatonin peak.

According to Lucas et al. [2014], lighting for shift work is an example of a situation when lighting design needs a balance between maximizing and minimizing non-visual responses:

- maximizing the non-visual response with bright lighting which will increase job performance, raise alertness, and decrease the risk for occupational accidents, but will simultaneously suppress melatonin and shift the circadian clock to an undesirable phase. In practice, retinal irradiance should be increased (but within acceptable blue light safety limits) and the spectral power distribution of light sources should be shifted to blue/green light regions (shorter wavelengths),
- minimizing the non-visual response by dim lighting which will minimize the shift of the circadian phase, but simultaneously decrease job performance, lower alertness, and increase the risk for occupational accidents. In practice retinal irradiance should be as low as possible and the spectral power distribution of light sources should be shifted to the red light region (longer visible wavelengths).

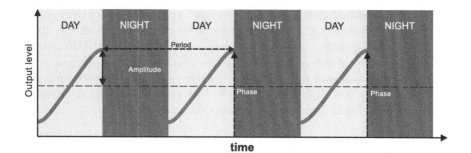

FIGURE 8.2   Circadian rhythm phase, amplitude, and period.

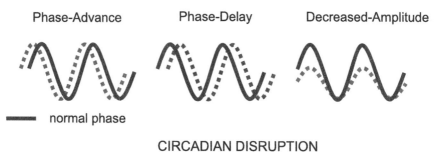

FIGURE 8.3   Presentation of phase advance, phase delay, and decreased amplitude.

8.2.2   Lighting Protocols Supporting the Occupant's Activity

Different work schedules in which employees rotate shifts are used, e.g. short-term (rotating) night shifts, and long-term (permanent) night shifts. There is a distinction between long-term (permanent) night shift work and varied (rotating) shift work regarding their circadian disruption and related lighting strategy.

In the available publications, several lighting scenarios for shift work could be found. They are based on various results of scientific studies, and therefore the proposed dynamic lighting protocols are different. Some selected lighting scenarios that can be an inspiration for lighting designers were chosen to be presented in this book. However, it should be remembered that along with the dynamic development of knowledge related to non-visual effects of light, the solutions presented can be modified in the future and the final lighting system should also take into account the individual characteristics of the user (age, chronotype, light exposure history etc.). However, it is important to be careful and remember that "avoidance of light at the wrong time is possibly more important than the light treatment itself" [Arendt 2010].

8.2.2.1   Short-Term (Rotating) Night Shift Work

Workers involved in rotating shift work are subjected to continuous stress in order to adjust as quickly as possible to the variable duty periods. In short-term night shift workers, shifting the sleep pattern is not required. It is recommended to maintain the day time circadian phase. There are several guidelines of lighting strategies focusing on the **no phase shift:**

1) using caffeine as an alerting stimulant and sleeping at home in a quiet dark place to preserve sleep and performance, or using goggles blocking the short wavelengths [Arendt 2010],
2) using neutral or warm white light and high illumination levels as an alerting stimulus at the beginning of shift work [Harmon 2019],
3) using selectively filtered light in the range below 480 nm to reduce sleep disruption [Rahman et al. 2013] and to maintain or even improve cognitive performance [Rahman et al. 2011; Rahman et al. 2013; Hoffman et al. 2008a; Kayumov et al. 2005]. It is important to note that this strategy has a major disadvantage, which is the impairment of color vision. The whole environment will be perceived as yellowish, while blue surfaces will appear black or green,
4) using constant neutral white light of 4000 K during all shifts and constant horizontal illuminance of 1000 lx during the evening and night shift until 2am, as presented in Figure 8.4,
5) using constant warm white 3000 K or neutral white 4000 K light and constant illuminances: horizontal – 300 lx, vertical at the eye – 100 lx (it corresponds to *MEDI* ≈ 50 lx) [van Bommel 2019b],
6) remaining entrained to the day shift and still being alert during the night shift [Figueiro and Hunter 2016a; LHA 2019]:
   - starting to use illuminance at the eye of no more than 30 lx at around 10pm,
   - during the dim light period, providing an alerting stimulus by using red (about 640 nm) light of at least 40–60 lx at the eye in rest areas or workspaces (see Figure 8.5),
   - exposure to red light can be used intermittently throughout the night,
   - if a high illuminance level on the task area is required, task lighting should be applied.

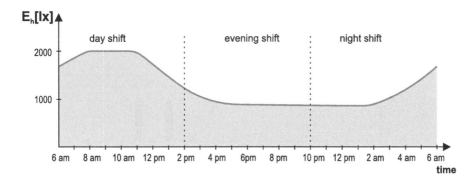

FIGURE 8.4 Example of lighting protocol for all shifts.

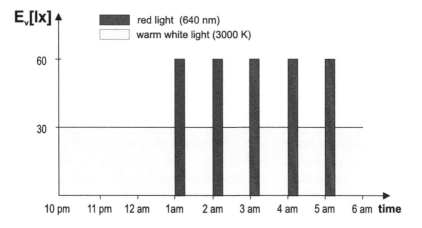

FIGURE 8.5   Example of lighting protocol for night shift with alerting red light exposure during the night

Guidelines of lighting strategies focusing on the **compromised-phase shift:**

1) remaining entrained to the day shift, but with adapting the evening chronotype "night owl" circadian rhythm [Figueiro and Hunter 2016a; LHA 2019]:
   - provide high illuminance of at least 200–300 lx at the eye until 3–4am
     and then dim light of illuminance of no more than 30 lx (see Figure 8.6),
   - during the dim-light period, provide an alerting stimulus by using red
     (640 nm) light of at least 40–60 lx at the eye in rest areas or workspaces,
   - exposure to red light can be used intermittently throughout the night,
   - if high illuminance level on the task area is required, task lighting
     should be applied,
   - use dark sunglasses on the way home after work to avoid outdoor light-
     ing which could affect night owl circadian rhythm adaptation.

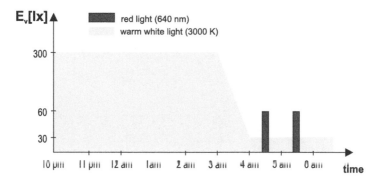

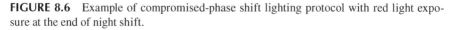

FIGURE 8.6   Example of compromised-phase shift lighting protocol with red light exposure at the end of night shift.

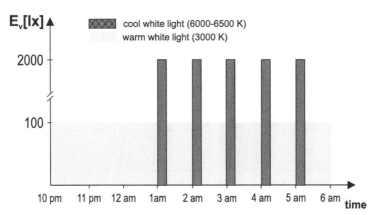

FIGURE 8.7 Example of compromised-phase shift lighting protocol with exposure to high intensity cool light after midnight (based on van Bommel 2019b, with permission).

2) maintain the overall ambient light at low levels during night shifts and provide additional bright light using light goggles or task lighting [Figueiro and Hunter 2016a; LHA 2019],

3) use constant warm white light of 3000 K and vertical illuminance at the eye of 100 lx until 1am. Then intermittently switch on an intense cool white light of 6500 K and vertical illuminance up to 3000 lx in cycles: 15 minutes on/45–60 minutes off [Van Bommel 2019] (see Figure 8.7).

8.2.2.2 Rotating "Swing Shift" Night Work

Among rotating night shift work it is worth mentioning 12 h shifts, or the so-called "swing shifts", of the following schedule: seven night shifts, 6pm–6am, then seven day shifts, 6am–6pm. For "swing shifts", the following schedules for day shift and night shift lighting is recommended [Glamox 2019]:

Day shift: dynamic lighting supporting occupant's activity (see Figure 8.8).

- The shift starts from warm light and a low level of cylindrical illuminance at the eye $E_{cyl} = 150$ lx (mimic morning sunrise), which is maintained for one or two of the first hours. Then activating cool light of 6500 K and $E_{cyl} = 350$ lx is switched on for 1–2 h. This way the cool white light shifts the daily rhythm forward.
- About 1pm there is a cool white shower to avoid the "post-lunch dip effect" and maintain performance.

Night shift (distinction between short-term night shifts depending on the number of days in a sequence):

a) short-term night shift (2–3 night shifts in a sequence) – employees do not need to shift the sleep pattern:

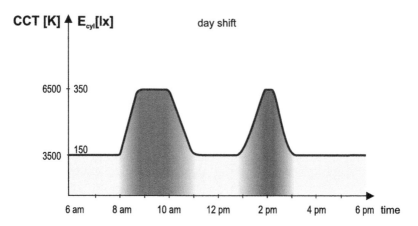

FIGURE 8.8   Figure 8.8. Example of day shift lighting protocol for "swing shift" with "morning boost" and a "post-lunch dip" increases in CCT and cylindrical illuminance.

- the shift starts from warm white light 3500 K and low cylindrical illuminance at the eye 150 lx and is maintained until 4am,
- activating cool light 6500 K of higher cylindrical illuminance 350 lx is switched on at 4am and maintained up to 7am (alternatively, alerting cool white "showers" every hour in the second part of the night shift could be applied)
- warm light 3500 K of lower illuminance 150 lx (the same parameters as at the beginning of the shift) is switched on at 7am and maintained to the end of the shift (see Figure 8.9).

b) short-term night shift (more than three night shifts in a sequence) – a temporary phase delay of about 8 h is recommended. Then the worker is awake at night and sleeps directly after the night shift. This kind of shift of the

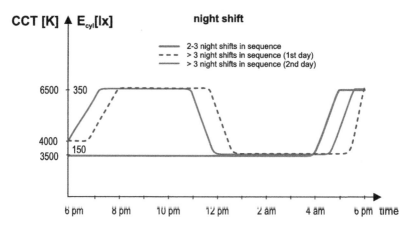

FIGURE 8.9   Night shift lighting protocol for short-term night "swing shift" (based on the idea presented in Glamox 2002)

circadian phase takes 4 to 5 days. The following is the lighting schedule for the first day:

- the shift starts from neutral white light and low cylindrical illuminance at eye level of about 250 lx and then illuminance and color temperature are gradually increased until 7am; then cool light 6500 K and illuminance 350 lx are maintained for 3–4 h to delay the phase,
- at about midnight, warm light 3500 K of lower cylindrical illuminance 150 lx is switched on and maintained for about 4 h, until around 4am,
- just after 4am, cool light 6500 K of cylindrical illuminance 350 lx is turned on again and maintained until 6am, the end of the shift (see Figure 8.9),
- on the second day, the activating cool light is switched on one hour later, i.e. at 8pm and left until midnight. This procedure is repeated until the phase is delayed by about 8 h.

8.2.2.3   Long-Term (Permanent) Night Shift Work

In the case of long-term, regularly scheduled night shift work, two different lighting scenarios could be found in the practical recommendations:

- no-inverted sleep–wake cycle – during the first half of the shift use high illumination levels and cool white light (6500 K) with a high proportion of blue light from the 460–480 nm wavelengths range to promote alertness as well as adaptation of the circadian rhythm to the long-term regularly scheduled shift work [Harmon, 2019],
- inverted sleep–wake cycle – during the first half of the shift use high illuminances: horizontal > 1200 lx, vertical at the eye 400 lx (it corresponds to 400 *EML* – equivalent melanopic lux) and cool white light (6500 K); sleep in total darkness in the morning after a night shift is a prerequisite for reversing the sleep–wake cycle; additionally it is recommended to minimize the amount of light reaching the retina in the period after work (wearing dark glasses on the way home and limiting exposure to other light sources to lower illumination levels and warmer light). [Harmon 2019; van Bommel 2019b]. It worth mentioning that night shift workers may almost completely adjust to the inverted sleep–wake cycle also on their days off [Folkard 2008; Costa 2010; Arendt 2010].

Many scientific studies have proven that blue light or white light with a significant proportion of blue light has a strong impact on night-time working performance, subjective feelings of alertness, and circadian physiology. This fact encouraged the use of this type of lighting as a stimulus for alertness during shift work. One of the first studies by Canazei et al. [2016] demonstrated that by using white polychromatic light with reduced short wavelengths (in the blue region), it is possible to both fulfill lighting standard requirements for indoor illumination and simultaneously make a positive impact on the physiology of night shift workers without detrimental consequences for cognitive performance and alertness. These results indicated that the use of white light, which has a small proportion of blue light, can be a compromise solution for healthy lighting, ensuring compliance with the standard requirements at night shifts. However, this solution may not satisfy the demands of cognitive performance and alertness maintenance during the second part of the night shift. Knowing that the use of cool white light (with a significant proportion of blue light) at the wrong time and with the wrong duration can result in worse results than not using it, the idea of introducing warm white light (which has a relatively small proportion of blue light) seems to be safer for workers' health. Studies, mainly carried out by researchers from the Lighting Research Center at Rensselaer Polytechnic Institute (LRC), proved that the use of red light at night can be as effective as blue light in increasing the level of alertness, without inhibiting the suppression of melatonin [Figueiro et al. 2009; Figueiro et al. 2010; Figueiro et al. 2016c; Figueiro et al. 2019b; Sahin et al. 2013]. Such outcomes showed another way of activating workers by light during the night shift – using red light as an activating stimulant.

8.2.2.4 Circadian Stimulus in Designing Lighting for Shift Work

Recommendations for designing dynamic lighting with a circadian stimulus for office work came first, shortly after verifying the model of phototransduction by the human circadian system in field studies at various workplaces [Rea et al. 2005; Rea et al. 2012; Rea et al. 2016]. However, recently (in 2019) researchers from LRC proposed using *CS* to design lighting for shift work. Basically, the idea of lighting for night shift work has not significantly changed compared to the approach presented in Chapter 7, but it seems important to indicate that *CS* also appeared here. Lighting protocols both for day shift and night shift work are presented below [LRC 2019a]:

- day shift lighting supporting the occupant's activity protocol:
  - $CS \geq 0.3$ throughout the day, or for a minimum of the first 2–3 h of the morning,
  - for maintaining alertness use $CS \geq 0.3$ throughout the day, or use red light as supplemental lighting during the afternoon.
- night shift lighting supporting alertness and performance – two protocols:
  - *phase shifting of the biological clock*, dedicated to permanent night shift work:
    - provide $CS \geq 0.3$ until 5am,
    - after 5am wear orange-tinted goggles to avoid circadian-effective light both to the end of the shift and on the way home,
    - sleep in a completely darkened room and avoid exposure to light until the afternoon.
  - *light therapy* – no phase shifting of the biological clock, dedicated to rotating night shift work:
    - ambient light of $CS = 0.1$,
    - task lighting to provide adequate illuminance for task performance,
    - red light doses (therapy) to increase alertness – in restricted areas.

8.2.3 Practical Examples of Shift Work Lighting

8.2.3.1 Desktop Luminaire

An interesting example of a desktop luminaire (developed by LRC) to provide an adequate circadian stimulus or an alerting stimulus is a 20 plug-in LED, which delivers three lighting modes throughout the day: blue (λ_{max} = 455 nm), white (6500 K), and red (λ_{max} = 634 nm). The luminaire is programmed in such a way that it automatically changes the color of light during the workday [LRC 2018a; LRC 2018b]. This luminaire was tested and verified in a few field studies. One of these was carried out in three office buildings in the USA and lighting was automatically changed according to the following protocol: blue light (*CS* = 0.4) in the morning (6am to 12pm), cool white light (*CS* = 0.2) at midday (12pm to 1:30pm), and red light (*CS* = 0) in the afternoon (1:30pm to 5pm). The lighting intervention was aimed at delivering a high circadian stimulus in the morning to promote entrainment and better quality of sleep, and red light in the afternoon to promote alertness. The lighting intervention (results based on three days of intervention only) produced a modest phase advance (earlier bedtimes and earlier waketimes) and better alertness in the afternoon [LRC 2018b; Pedler et al. 2017; Figueiro and Hunter 2016a]. This desktop luminaire could also be programmed for different shifts. It could be used both during daytime shift work and during night shift work. During night shift work with phase shifting of the biological clock, the cool white light mode can be used, while the red light mode can be used during night shifts with no phase shift (therapy). However, during the day shift all three modes can be used: blue and/or white light (depending on the visual task requirements, e.g. blue light could be avoided because of color rendering needs) in the morning until 12am or even longer, and red light in the afternoon.

The construction of this lamp is mainly dedicated to placing it on the desk in such a way that it is below the lower edge of the monitor (or behind the laptop screen) in front of the employee, so that it is always in the field of view and the light falling on the eye is partly within the range of a good photobiological effect and partly within the range relevant for the visual task. Considering the above, this luminaire can very well fulfill its role at stationary shift workstations, where the employee performs activities in a fixed, stationary position, e.g. in control rooms and call centers. However, at a shift work workstation where the worker constantly moves, even within one workstation, its applicability in terms of effectiveness is questionable. In addition, the spatial layout of the workstation, its equipment, and technological processes may be a significant obstacle in its application or installation.

The question, therefore, is how to solve the problem of supplemental lighting which will be suitable for a shift worker moving within his workstation during the performance of work, when it is not possible to install a desktop or standing luminaire. Perhaps using a localized luminaire will prove to be a solution.

8.2.3.2 Localized Luminaire

Researchers from the Central Institute for Labour Protection – National Research Institute (Poland) together with GL Optic Poland have developed and built a model of a tunable LED-based localized luminaire for dynamic lighting to provide an adequate circadian or alerting stimulus at shift work workstations in an automotive industry company. Assumptions for building the model took into account the work characteristics and the limitations related to the work performed and the technological process. The employees performed work in a standing position while moving around the task area – a car engine assembly workplace.

At the time of writing this book, the results of the pilot field study at the test workstation equipped with a model of the luminaire have just been finished, so a description of this solution is not yet included in any available publication. For this reason, a slightly more detailed description of this solution is provided here.

In order to obtain different spectral distributions of emitted light (including saturated red and blue) and the corresponding different color temperatures of white light and at least a good color rendering index $Ra > 90$, $R9 > 50$, additional diodes with blue 460 nm, royal blue 445 nm, green 520 nm, cyan 505 nm, and red 630 nm were applied. The luminaire is fully programmable and allows the creation of any different lighting scenarios (sets of lighting scenes of different illuminance and SPD) within 24 h. The luminous flux control system enables both "smooth" changes in spectral power distribution (and associated changes in light color (CCT)) and smooth changes in illuminance at a time interval from 1 minute to 24 h. Each of the programmed light scenarios is saved with a name assigned to it and can be called up by the user at any time. The program automatically sets the light scene of the scenario that falls on the hour at which the scene was called. The illuminance produced by the luminaire ensures red light (630 nm) and blue light (460 nm) at the eye on a vertical plane (at the height of 1.6 m – standing position) in the range up to 40 lx. The views of old localized lighting and new localized lighting (installed model of localized luminaire) are presented in Figure 8.10. The

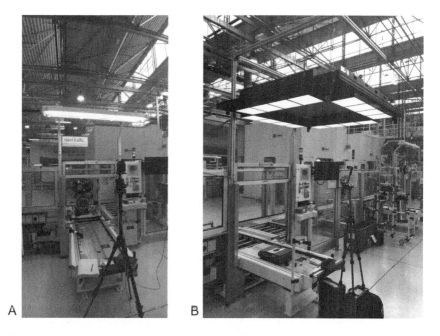

FIGURE 8.10   View of the shift workstation with: (A) old localized lighting; (B) new localized lighting designed for the study.

old localized lighting was a fluorescent luminaire 2×58 W with diffuser cover. At the worker's eye it provided $CS = 0.43$, $M/P = 0.61$ (measured discomfort glare: $UGR = 24$ – see Figure 4.2a in Section 4.3.2), regardless of the work shift (the luminaire had no control of luminous flux output and was not dimmable).

As can be seen in Figure 8.10b, the luminaire consists of 16 LED modules (with dedicated diffusing optics) connected to each other in the shape of a square. They are arranged in such a way that the task area can be evenly illuminated and not shaded by a worker standing in different places around the engine. Discomfort glare was adequately limited ($UGR = 18$ – see Figure 4.2b in Section 4.3.2). In order to eliminate the influence of general lighting in the factory hall, the inner part of the square of the luminaire was obscured by light-absorbing material. The view of the workplace with an installed model of the localized luminaire during a worker's activities is presented in Figure 8.11.

The single LED panel has luminous intensity distribution close to Lambertian with the maximum output 498 lm (see Figure 8.12 in Table 8.1). Possible values of CS determined from the field measurements at the test workstation (which includes reflection from the environment) while maintaining $Ra > 80$ and horizontal illuminance of at least 500 lx (according to standard EN 12464-1 requirement) were in the range of 0.17 to 0.67. In addition, it is possible to switch on blue, green, cyan, or red light only. For example, red light can be applied during the night for resting periods [LRC 2019b] or can be used intermittently [Figueiro and Hunter 2016a].

The results of simulation calculations for the configuration of 16 LED modules used in the luminaire model layout with an assumption of 250 lx and 300 lx vertical at the eye are presented in Table 8.1. The summary of results is presented in the Table in the same arrangement as in the article by Jarboe et al. [2019], so that it can be compared with published simulations for other luminaires. The obtained ratio $E_v/E_h = 0.65$ meets the condition of obtaining $CS = 0.3$, while maintaining a relatively low illuminance on a horizontal working plane [Jarboe et al. 2019].

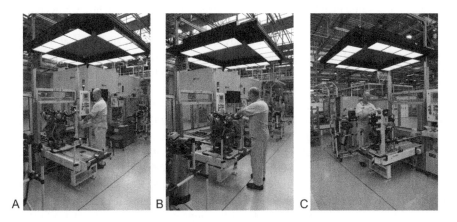

FIGURE 8.11 View of the workplace with an installed model of the localized luminaire during a worker's activities.

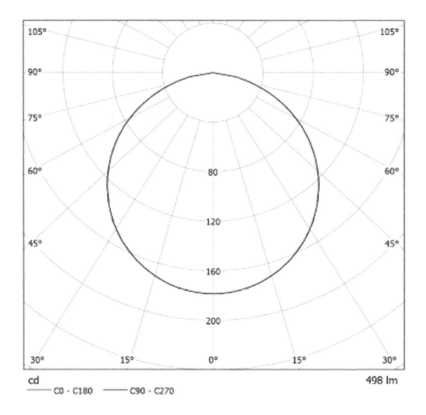

FIGURE 8.12 Luminous intensity distribution of single LED panel (20 W, 498 lm) of localized luminaire presented in Figure 8.10B and Figure 8.11.

The results presented in Table 8.1 are consistent with current knowledge. As CCT increases, the value of *EML* and melanopic irradiance increases, while *CS* decreases at CCT = 3500 K compared to 3000 K and only reaches 0.3 at CCT = 5500 K for E_v = 250 lx, while at E_v = 300 lx already at CCT = 3000 K and next at 4500 K. However, if we look at the *EML* results, it can be seen that the level of *EML* = 150 is already obtained at CCT = 3500 K for E_v = 250 lx, and at 3000 K for E_v = 300 lx.

Considering the requirements of the WELL building standard and the LRC recommendations, the results obtained give different guidelines to designers regarding the use of both light color and minimum illuminance at the eyes. By adjusting to the WELL standard and using a warm white light color with CCT = 3500 K, we can already get *EML* = 150 with E_v = 250 lx, and already with CCT = 3000 K (and each next higher CCT) with E_v = 300 lx. However, considering the LRC recommendations to use only cool white light color with CCT ≥ 5000 K, we can get *CS* = 0.3 at E_v = 250 lx, while with E_v = 300 lx, we can also get that value at CCT = 3000 K and next at 4500 K (see Table 8.1). This shows the impact of the adopted metric of effective biological lighting on the lighting design and its subsequent implementation in a real working environment.

TABLE 8.1

Results of Photometric, Circadian, and α-Opic Irradiances for a Luminaire Model (of 16 (4x4) LED Modules; $E_v/E_h = 0.65$)

Parameter	250 lx vertical (simulation)					
CCT	3000	3500	4000	4500	5000	5500
E_h	385	386	388	389	390	391
E_v	250	251	252	253	253	254
CLA	256	123	180	231	268	310
CS	0.29	0.17	0.22	0.27	**0.30**	**0.32**
ipRGC*	16.87	20.29	23.38	26.13	28.31	30.38
Rod*	21.44	24.60	27.39	29.86	31.91	33.67
S-cone*	6.97	9.37	11.70	13.79	15.28	17.06
M-cone*	29.69	31.29	32.62	33.77	34.70	35.45
L-cone*	41.29	41.15	41.15	41.25	41.18	41.25
EML	140	**169**	**194**	**217**	**235**	**253**
M/P	0.56	0.67	0.77	0.86	0.93	0.99
Parameter	**300 lx vertical (simulation)**					
CCT	3000	3500	4000	4500	5000	5500
E_h	461	462	464	464	465	465
E_v	300	300	302	302	303	303
CLA	307	149	217	278	321	371
CS	**0.32**	0.19	0.26	**0.30**	**0.33**	**0.36**
ipRGC*	20.20	24.34	28.06	31.24	33.81	36.23
Rod*	25.68	29.51	32.88	35.69	38.12	40.17
S-cone*	8.40	11.27	14.05	16.55	18.28	20.37
M-cone*	35.54	37.48	39.08	40.31	41.45	42.27
L-cone*	49.37	49.23	49.24	49.19	49.16	49.14
EML	**168**	**202**	**233**	**260**	**281**	**301**
M/P	0.56	0.67	0.77	0.86	0.93	1.00

E_h – average horizontal illuminance, E_v – average vertical (at the eye cornea) illuminance, CLA – average circadian light, CS – average circadian stimulus (calculated using CS Calculator [2019]), *CIE α-opic responses for ipRGC (melanopsin), rods, cones (long – L, M – medium, and S – short) calculated using CIE DIS 026/E:2018 [CIE 2018] equations and α-opic action spectra, EML – equivalent melanopic lux, M/P – melanopic ratio (melanopic irradiance/photopic illuminance ratio) (calculated using WELL calculator [2019]).

After installing the luminaire model that was developed at the workplace, a series of spectroradiometric measurements were made to adapt the emitted luminous flux at different spectral distributions of light to the adopted assumptions of light scenarios for each work shift. In this way, the luminaire was calibrated taking into account the component reflected from the surroundings which reaches the worker's eyes. The determined melanopic ratios (M/P) depending on the CCT color temperatures determined at the worker's eyes are presented in the upper panel of Figure 8.13. At CCT

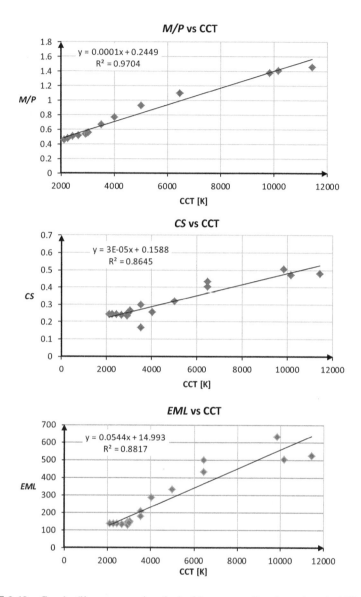

FIGURE 8.13   Graphs (linear approximation) of (upper panel) melanopic ratio *M/P*, (middle panel) *CS*, and (lower panel) *EML* for different CCTs.

6500 K color temperature, the melanopic ratio *M/P* was 1.1, which means that the proportion of melanopic irradiance is the same as for daylight with the same color temperature [Brennan et al. 2018]. It is worth knowing that in daylight (sky spectra) with a color temperature of 6500 K, *CS* = 0.3 and *EML* = 200 can be obtained at an illuminance of about 182 lx [Brennan et al. 2018]. For cool white light spectra with CCT 6500 K of our model of localized luminaire, the *CS* = 0.3 and *EML* = 205 can be obtained at an illuminance of about 186 lx at the worker's eye (measured at

the workplace by increasing the vertical illuminance at the eye to about 320 lx, the $CS = 0.407$ (Figure 8.13, middle panel), and $EML = 430$ (Figure 8.13, lower panel). Still, we should remember that the same CCT could get different M/P, CS, and EML depending on the spectral distribution of light.

Subsequently, those modeled scenes were selected whose horizontal illuminance E_h and CS values were in the recommended value ranges and provided $Ra > 80$. As it turned out in practice, after installing the model at the workplace, it was not possible to obtain $CS < 0.17$ while maintaining $Ra > 80$ and E_h about 500 lx. This could also be influenced by both the direct luminous intensity distribution (not recommended for use [Figueiro et al. 2016b]); however, the E_v/E_h ratio was 0.65, which is in line with Jarboe et al. [Jarboe at al. 2019] and the indirect component from light reflected from the surroundings. Considering the above, we decided to use the following night shift protocol: to maintain $CS \approx 0.24$; throughout the night shift (even if this CS value is not recommended for night shift work), but instead adding an alerting red light to the white light spectrum in proportions which did not cause the color rendering to be less than $Ra = 80$.

The lighting protocols for particular work shifts were as follows:

- day shift lighting protocol was based on the lighting schedule proposed by Figuerio et al. [2016b], the shift started at 6am from cool white light 6500 K and vertical illuminance at the eye of 350 lx which corresponds to $CS = 0.43$ (it is worth mentioning that with "old lighting", all working shifts had a constant CS value which was 0.43 at CCT = 4000 K). Then illuminance and CCT were gradually decreased until they reached 5000 K and 280 lx ($CS = 0.32$) at 8am; between 8am and 12pm those parameters were maintained; between 12pm and 1pm, CCT was gradually decreased to 4000 K; from 1pm to 2pm, the CCT and illuminance achieved respectively 3500 K and 245 lx ($CS = 0.17$) (see Figure 8.14A).
- evening shift lighting – (between 2pm and 6pm the lighting protocol was based on the schedule proposed by Figueiro et al. [2016b]): the shift started at 2pm from warm white light 3500 K and vertical illuminance at the eye 245 lx ($CS = 0.17$); between 2pm and 6pm CCT and illuminance were gradually decreased to 3000 K and 210 lx ($CS = 0.2$); between 6pm and 8pm CCT was gradually decreased to 2900 K, and between 8pm and 10pm illuminance decreased to 290 lx ($CS = 0.23$) (see Figure 8.14B);
- night shift lighting – the shift started at 10pm from warm white light 2900 K and vertical illuminance at the eye 190 lx ($CS \approx 0.24$); between 10pm and 11pm CCT and illuminance were maintained; after 11pm until 3am, 10 lx of red light was gradually added every hour (CCT ranges between 2611 K and 2064 K, $CS \approx 0.24$); after 4am, the red light component was gradually taken away; and at 6am 2900 K and 190 lx was achieved (throughout the shift $CS \approx 0.24 \approx$ const) (see Figure 8.14C).

The 6-month-long field study was carried out at two shift work workstations performing the same work task: one with an installed localized luminaire model (new lighting) and the other with unchanged lighting (old lighting, fluorescent luminaire 2 × 36 W, 4000 K, $Ra > 80$ – see Figure 8.10a).

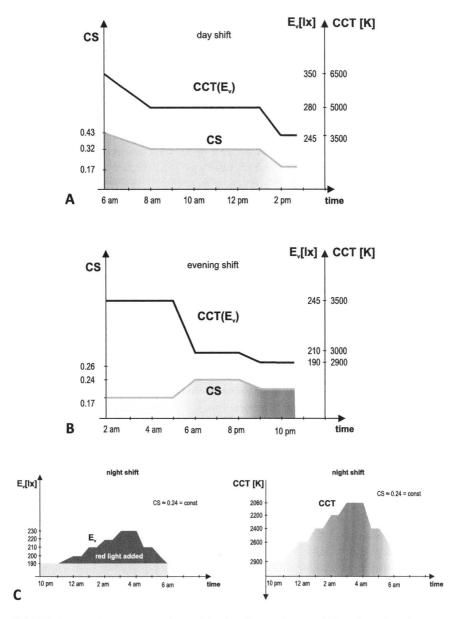

FIGURE 8.14 Lighting protocols used in the pilot study at a shiftwork workstation: (A) day shift; (B) evening shift; (C) night shift

Nine shift workers took part in the study, including three from each of the three shift crews (work schedule: 6 days' work/1 day off). Each of them took part in the study protocol for 54 work days (18 work days per each shift) twice a day: at the beginning and at the end of the work shift. The study protocol consisted of Grandjean's scale of fatigue (it is a visual analog scale VAS: 0–100 mm), subjective sleepiness evaluation using the Karolinska Sleepiness Scale (KSS) [Akerstedt et al. 1990], a psychophysical performance test – the Line Judgment test (from the PEBL test battery developed by Esposito et al. [2011]), and subjective assessments of lighting in terms of comfort, well-being, and lighting influence on mood (a questionnaire especially developed by CIOP researchers for the needs of this study). The assessment of the lighting influence on workers' mood consisted of choosing one of the two opposing terms from among the five paired sets: relaxing – irritating, pleasant – unpleasant, stimulating – sleepy, appropriately bright – too bright, favorable to work – disturbing work.

Some of the results are presented below. The Grandjean subjective indicator of fatigue consists of the following subscales: strong – weak, rested – tired, interested – bored, vigorous – exhausted, awake – sleepy, effective in action – ineffective in action, attentive – distracted, and able to concentrate – unable to concentrate. The highest value on the scale was the highest indicator of fatigue. The calculated values of assessment differences "after–before" the shift (the mean answers of all the workers who volunteered for the pilot study) for both lighting conditions – old lighting and new lighting – and for each type of shift work are presented in Figure 8.15. A higher value of the difference means greater fatigue after the shift. The results indicate that on day shifts employees were more tired at the end of work when they were using the new lighting than the old, whereas on the evening and night shifts, the opposite was true; they were more tired after using the old lighting than the new. This may suggest that the lighting protocol with the assigned schedule of changes: *CS*, CCT, and illuminance (developed by Figueiro et al. 2016b) intended for daytime office work is not as good for day shift work. The study should be repeated for more reliable results and conclusions. It is interesting, however, that the results indicated the occurrence of smaller differences in fatigue after the evening and night shifts with new lighting. This would mean workers were less fatigued when using the new lighting than the old.

Regardless of the lighting conditions, subjectively assessed alertness remained at practically the same level before and at the end of the work shift (they did not differ significantly), especially on day and evening shifts. On the night shift, alertness decreased after work on average by one unit on the scale from alert to fairly alert on the KSS scale for both lighting conditions, which means that more bluish lighting (old lighting) and more reddish lighting (new lighting) produced the same effect. It is worth noting that on night shifts employees were sleepier at the beginning of the work shift (more than at the end of the morning and evening shift), which could be related to their chronotype: morning and undifferentiated. One way or another, the type of lighting did not have a significant impact on the subjective assessment of employees' sleepiness. The average values of the subjective alertness assessment are presented in Figure 8.16.

In the psychophysical performance test (the Line Judgment test), the results obtained in terms of accuracy did not significantly differ for either type of lighting, but it can be seen that regardless of the type of lighting, the accuracy of completing the task did not significantly differ before and at the end of the shift work. In contrast to accuracy, the type of lighting significantly affected the response time of the subjects (all shifts analyzed together). The difference in mean response time between old and new lighting for all the shifts together are presented in Figure 8.17.

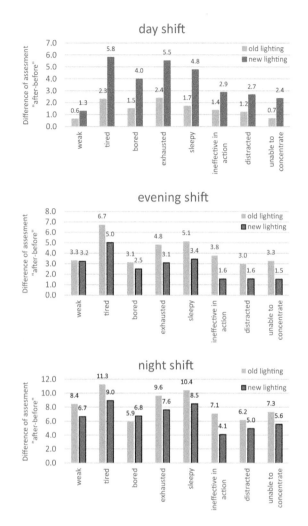

FIGURE 8.15 Differences between "before–after" fatigue assessment using Grandjean's fatigue scale for old and new lighting: (upper panel) for day shift; (middle panel) for evening shift; (lower panel) for night shift.

It was significantly shorter for new lighting, which indicates better performance. In the morning shift, both the old and new lighting shows a decrease in response time. In the evening shift, the old lighting shows an increase in response time, while the new lighting shows a decrease in response time. Such results could be due to the fact that during the night shift, the old lighting provided a high circadian stimulus ($CS = 0.43$), while the new lighting provided a smaller one – almost half of the old lighting ($CS \approx 0.24$). On night shifts with both types of lighting there was a practically steady response time and accuracy. However, one should take into account the fact that keeping the response time at a constant level during the night shift with new lighting indicates the practicality of this solution.

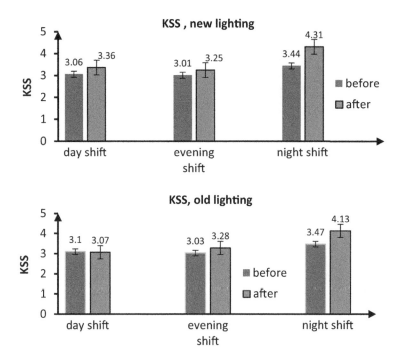

FIGURE 8.16 Mean KSS values (higher KSS value – lower alertness level) for all the workers surveyed, obtained at each shift before and after work: (a) while working under new lighting; (b) while working under old lighting.

All the workers who took part in the pilot study assessed new lighting as pleasant, stimulating, appropriately bright, and favorable to work. Research results have shown that employees rated the new lighting much better than the old. The warm light color at night was considered more pleasant than the previous neutral white, more favorable to work, and more relaxing.

Although the pilot study did not show a statistically significant improvement in alertness and psychomotor test results, the well-being of employees associated with working when using the localized lamp model that was developed is important. We also have to remember that maintenance of work performance, especially during the night shift, is a very good result. Nevertheless, some improvements in the lighting scheme for each shift could be made. For example, on night shifts it would be worth considering the use of an intermittent red light addition to the white light spectrum, e.g. at 15-minute intervals (as a kind of "light shower") or, ultimately, saturated red light to get a better effect on alertness and performance. These results are only an encouraging introduction to subsequent studies, with changed lighting scenarios, so as to achieve maintenance of work performance on the one hand, and healthy lighting on the other.

8.2.3.3 General Lighting

The idea of new circadian lighting solutions for use as general lighting was introduced by Circadian Light, which developed a new night LED based on a violet LED

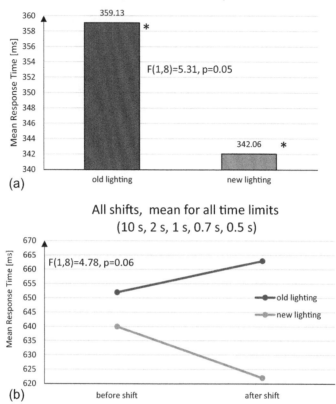

FIGURE 8.17 Mean response time in the Line Judgment test for old and new lighting (a) for all shifts and 0.5s limit time for response; (b) for all shifts before and after shift and mean of all limits of time for response (10 s, 2 s, 1 s, 0.7 s, 0.5 s).

of maximum emission at the wavelength of 405 nm and dedicated phosphor, which was launched on the market. This LED idea has its origin in the medical research of the scientific team led by Martin Moore-Ede. At the same time, it fits into the new circadian metric he proposed, i.e. circadian potency (see Chapter 6). Since light sources should practically give out no blue component at night in order to avoid circadian disruption, this LED emits only 2% blue light from the region between 440 nm and 490 nm compared to all the visible region. This solution makes it possible to obtain the required high illuminance level at the working plane together with a high quality of white light and not influence the circadian system. A new series of "circadian" luminaires with new LEDs have been developed. They are based on two types of LEDs: day LEDs which contains 20% blue light, and night LEDs with 2% blue light (compare Figure 8.18).

This new night LED solution has a good chance of being widely implemented, primarily due to the ease of its use for night shift lighting. Perhaps the end of the era of blue-pump LEDs is approaching, giving way to the new era of violet-pump LEDs.

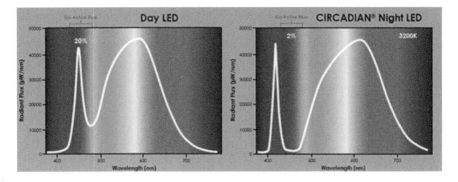

FIGURE 8.18   Spectral distribution of day LED and circadian night LED for shift work general lighting (source: Circadian ZircLight 2019b, with permission).

8.3   SUMMARY

When trying to cope with circadian disruption, we use lighting for better circadian adaptation to night shift work. Thanks to this, workers are supposed to achieve better performance, alertness, and mood, as well as longer and better daytime sleep. To obtain this goal, various shift work lighting schedules and guidelines have been described in this chapter, which provide a fairly wide range of choices for the lighting designer. It should be remembered, however, that we design lighting for people whose psychophysical possibilities differ. The adaptation to night shift work is associated with individual susceptibility to circadian misalignment, which should be considered both for permanent night shifts and for rotating shift work [Boudreau et al. 2013].

Localized lighting with programmable light scenes at shift workstations seems to be a good solution wherever local desktop lighting is impossible to use. It also gives the opportunity to adapt to the individual needs and characteristics of the employee, which is part of the idea of integrative (human-centric) lighting. The statement made by Jarboe et al. [2019] that the strategy with the most impact is general lighting and desktop lighting delivering light directly to the worker's eyes will not be effective in workplaces where the worker moves and does not have a stationary observation position, for example, at office workstations. Moreover, localized luminaire lighting also allows the use of short light showers, e.g. red light during a night shift. Switching on red light can also be a signal that it is time for a break from work and for going to a room for socializing, which is provided with adequate lighting. The use of localized lighting makes it possible to meet the recommendations regarding healthy lighting supporting alertness and performance both for rotating and permanent shift work, which were presented in this chapter.

The type of shift work (rotating, permanent, swing etc.), user preferences, and characteristics – are crucial at the beginning of the design process. It is advisable to agree on design assumptions with worker representatives and the local health management authorities responsible for occupational health and safety.

In addition to substantive issues related to visual and non-visual effects of light, designers must also take into account customer requirements, installation costs, and energy consumption. However, including these should not prevent them from creating healthy lighting. It is a particular challenge to design human-centric lighting for night shift work.

9 Human-Centric/ Integrative Lighting Design

As we work to create light for others, we naturally light our own way.

Mary Anne Radmacher

9.1 THE PHILOSOPHY OF HUMAN-CENTRIC LIGHTING

Human-centric lighting (HCL) philosophy is related to the mission of ergonomics. Accordingly, human health and well-being are of foremost importance in lighting design. A holistic approach and catering for human biological, emotional, and visual needs have high potential that can benefit people and bear fruit in long-term health, well-being, and performance.

In 2016 the International Committee on Illumination, CIE, introduced a new term – integrative lighting, the definition of which is: "lighting that is specifically designed to produce a beneficial physiological and/or psychological effect upon humans" [CIE 2016a]. Some other terms, such as "human-centric lighting" (HCL), "circadian lighting", and "biodynamic lighting" can be treated as synonymous with integrative lighting.

The three pillars of HCL are biological, emotional, and visual light effects. They should be considered in the course of designing, commissioning, and operating when providing human-centric lighting, as presented in Figure 9.1.

Visual needs relate to meeting normative requirements, such as EN 12464-1 [2011] for all the lighting parameters listed therein, including specific lighting requirements for elderly people or people with visual impairments. Emotional needs relate to the well-being of humans within their social environment, which should provide a high level of acceptance, satisfaction, and well-being [licht.wissen, 21]. Biological needs relate to circadian rhythm entrainment, better sleep at night, higher attention, cognitive performance, and alertness during work.

Some earlier chapters (Chapters 4–8) describe the parameters affecting both visual and non-visual responses to light, together with examples of recommendations and practical design of lighting focused on the non-visual effects of light. Given that a number of existing publications describe lighting design in detail, including the visual effects of light (for example [Boyce 2014; van Bommel 2019a], and pay

FIGURE 9.1 Human-centric lighting (HCL): interdependence of visual, emotional, and biological effects of light on humans.

much less attention to a more detailed description of the non-visual effect, in this chapter we focus on the latter aspects.

9.2 GENERAL GUIDELINES

When starting the design of integrative lighting, it is necessary to predetermine five basic assumptions related to client needs: lighting requirements for anticipated visual tasks, architectural interior conditions, lighting control, and last, but not least, the expectations and characteristics of users. Those assumptions are listed below:

1) client needs (interior and workplace characteristics, work hours during the day (shift work), efficacy, aesthetic considerations),
2) use of light:
    • in the morning, delivery of the "boosting" light effect; in the evening and at night, no "boosting" and dimmed light,
    • the benefit of monochromatic light (blue – in the morning; red – in the post-lunch dip and at night),
    • including daylight or not,
    • general lighting only, adding task or localized lighting.
3) architectural finishes (diffuse or direct reflections, spectral reflectivity in the visual range, accent colors),
4) controlling the light: smart light scheduling, daylight harvesting, task tuning, occupancy control, personal control, variable load shedding, combined energy savings, high-tech, such as machine learning etc.,
5) occupant's characteristics (age, gender, chronotype) and expectations (planned lighting at the individual level and controllability),
6) occupant's training and information (explanation of the applied lighting control and human-centric lighting assumptions)

Having made these general assumptions, we can start to design, while bearing in mind both the visual and non-visual effects of light.

9.3 VISUAL EFFECTS OF LIGHT

9.3.1 Standard Lighting Code-Based Design

Lighting code design is about meeting the normative requirements and recommendations in the scope of a given interior and type of visual task. This applies to all the parameters dedicated to good quality of lighting listed in Table 4.1, Section 4.1.3, i.e. in accordance with EN 12464-1 [2011]: illuminance scale and uniformity of illuminance (task area, immediate surrounding area, background area, walls, ceiling), illuminance level measurement grid, luminance distribution, surface reflections (but not spectral ones), disability glare (shielding), discomfort glare (*UGR*), reflected glare, cylindrical illuminance, modeling, directional lighting, color aspects (CCT, color rendering index *Ra*), flickering and stroboscopic effects, illumination of computer workstations with additional requirements, the maintenance factor, energy efficiency requirements, daylight contribution, variability of light.

It should be emphasized at this point that designers should make sure that the minimum normative requirements are met while providing healthy lighting in the sense of catering for the non-visual effect of light.

9.3.2 Principles Beyond the Standard Lighting Codes

The development of knowledge in the field of lighting assessment also applies to parameters related to the visual effects of light which are not included in the lighting standards yet. It is worth knowing and remembering them, especially when working on lighting designs that use LEDs.

- The development of knowledge in the field of colorimetry and especially the establishment of new color-rendering indexes that better and more accurately determine the color rendering properties than those previously used (*CRI* or *Ra*) should be taken into account at the design level. Therefore, when choosing light sources, at least color fidelity R_f and color gamut R_g should be taken into account [Rossi 2019].
- It is worth implementing the method of selecting the illuminance level based on the parameters of the visual task and the related visual performance using Relative Visual Performance (*RVP*), especially when choosing universal illuminance for designing [Rea 2018] or when dealing with a special case of visual activity (not specified in the standard) (*RVP* is described in Chapter 5). It is worth taking this parameter into account, as it is very important for visual performance, especially in situations where we use circadian metrics (*CS* or *EML* or other metric) and adjust to each other the vertical illuminance at the eyes and the horizontal illuminance on the working plane.
- In interiors with glass walls or high daylight access through large windows, consideration should be given to the glare limitation caused by daylight. It is recommended to determine Daylight Glare Probability (*DGP*) and provide an appropriate window covering system.

- The development of knowledge in the field of flicker and stroboscopic effect evaluation make it possible to assess those phenomena by means of new metrics: flicker frequency, flicker percent, and flicker index, or stroboscopic visual measure *SVM* [CIE 2016c]. It is recommended that attention should be paid to LED flickering, especially when dimmed (more information in Chapter 10).

9.4   NON-VISUAL EFFECTS OF LIGHT

9.4.1   CIRCADIAN LIGHTING GUIDELINES

In principle, the selection of a circadian metric is crucial for this part of designing. It mainly affects the value of illuminance determined at the eye as well as the choice of light color or the use of different colors of light. Because of this, it is worth listing important aspects to be considered when designing circadian lighting:

- the value of simulated vertical illuminance at the eye depends on the chosen circadian metric; this could be: melanopic daylight equivalent (D65) illuminance, equivalent melanopic lux (*EML*), or illuminance determined on the basis of the established circadian stimulus (*CS*),
- simulation of vertical illuminance should be done from all light sources and should include reflected light,
- when choosing light sources, attention should be paid to the spectral power distribution of light source in order to achieve adequate circadian metric and both the desirable light color temperature (CCT) and the color rendering index,
- "tunable-white" or "RGB-tunable" [Dai et al. 2018] technology should be applied, because they are enabled to provide daily variation in melanopic metric (dimming and boosting illuminance, tuning the spectrum),
- timing and duration of different light scenes (boosting, alerting, and relaxing effects) should be applied according to the chosen circadian metric; the examples of lighting protocols are presented in Section 7.4 and for shift work in Section 8.2.2,
- daylight should be maximized and use made of cool white light in the morning; it is worth remembering that the circadian resetting period is between 6am and 10am; a limited effect on the resetting period occurs about noon, but at that time we can increase alertness [Andersen et al. 2012; Konis 2017],
- the field of view occupants have in relation to light sources and windows should be carefully adapted; it should be considered that at the workstations placed far away from the windows, adequate illuminance at the eye depends only on electric lighting; on the other hand, workstations placed close to the windows should be arranged so as to get as much illuminance as possible at the eye from daylight, but not cause glare from the windows,
- where applicable, contribution of light from computer monitors that can provide up to 25% of required *EML* (i.e. 50 *EML*) should be included [Brennan, 2018],

- where applicable, contribution of light from desktop lighting that can provide up to 20% of required *EML* (i.e. 40 *EML*) should be included [Brennan, 2018]. In addition, the use of task lighting affects the feeling of autonomy. It can result in higher job satisfaction and performance and decrease absenteeism [Juslen et al. 2007],
- glare from light sources (artificial and daylight) should be carefully considered, since a high level of light at the eye is required for circadian stimulation,
- the decrease in human lens transmittance with age when determining circadian metrics for occupants older or younger than 32 years old should be taken into account; use spectral correction factors given, for example, in CIE S 026 [CIE 2018]; (a 32-year-old standard observer was chosen for reference),
- exposure duration has a significant impact on suppressed melatonin – longer duration exposure suppresses melatonin to a greater degree [Nagare et al. 2019a; Nagare et al. 2019b].

The golden rule of circadian lighting is that it is not the light quantity, but the timing and dynamics of light that is critical to provide an effect on the circadian phase.

9.4.2 COGNITIVE AND CREATIVE PERFORMANCE VS SOME LIGHTING ASPECTS

As stated in Section 9.1, one of the pillars of HCL is biological needs. These are related to performance, especially cognitive performance.

Cognitive performance is the ability of an individual to carry out the various mental activities most closely associated with learning and problem solving. The speed of task solving and the accuracy of the solution are the measures of cognitive demands. Environmental factors can affect concentration, attention, and other cognitive processes and thereby impact speed and accuracy. A study concerning the aspect of lighting design concerning luminaire light distribution and its hypothetical influence on cognitive performance was carried out in Norway [Fostervold et al. 2008]. The results did not show any significant effect of light distribution on cognitive performance. This suggests that the choice of luminaire light distribution (direct, indirect, or direct–indirect) can significantly affect the luminous environment, the amount of the reflected component in the illuminance at the eye, and energy consumption, but not cognitive performance.

The other study of Tonetti et al. [2019] showed that even a single short exposure (1 min) to blue light (λ_{max} = 480 nm) during the post-lunch dip can improve cognitive performance in young adults (mean age 24 years) with a special emphasis on semantic priming and reorienting of attention, without modifying alertness.

In the literature we can also find the term *creative performance*, which is commonly assessed on the basis of the fluency, originality, and flexibility of performing particular tasks. The improvement of creativity may be caused by a variety of psychological factors, such as intelligence, attention, and mood. The latter could also be influenced by colors and light. The research of Kombeiz and Steidle showed that red and blue accent lighting on the wall in front of the occupant indirectly influences creative performance. Red accent lighting creates a friendly room atmosphere that may elicit motivation. Both blue and red light increased the strategic approach to motivation in relation to white light and indirectly improved creative performance [Kombeiz et al. 2018]. This implies that using color accents like blue and red accent lighting could indirectly influence creative performance. Such "decorative aspects for creativity" are worth being considered by architects, especially when designing for managers, and other workers whose professions demand creativity.

9.5　IMPORTANCE OF ROOM SURFACE REFLECTANCE AND COLOR

Some studies showed the importance of room surface reflectance and color in achieving vertical illuminance at eye level, all of which influence the circadian metric value [Dai et al. 2018; Bussato et al. 2018; Brennan et al. 2018; Konis 2017; Ewing et al. 2017; Bellia et al. 2017; Hartman et al. 2016].

The study by Dai et al. [2018] indicates that improving the reflectance of room surfaces and using indirect luminaires (the luminous flux directed upward) together with a high reflective ceiling can effectively improve both the efficacy of circadian lighting (up to 3.6 times) and the uniformity of spatial corneal illuminance (which is important for the circadian effect). Although their results are interesting and directly related to design practice, attention should be paid to two aspects: the possibility of discomfort glare from highly reflecting surfaces of the room and the adoption of absolute photopic reflectance, which is determined from the ratio of the reflected luminous flux to the incident luminous flux under the given conditions. Since reflectance is determined based on a photometric parameter which is weighted by photopic action spectra $V(\lambda)$, its applicability to determine circadian or melanopic metric is burdened with an error. More reliable results would be taking into account the spectral reflectance in the visible radiation range or absolute circadian reflectance (determined by using the same melanopic or circadian effectiveness function as the metric considered). Reflections from the surface of the environment should be assessed not only quantitatively but also qualitatively, taking into account the spectrum of reflected light. Let us assume there are two surfaces: red and blue, with the same reflectance. The reflected spectra from both surfaces would be different. With blue walls, the reflected light spectrum will be "enriched" in the blue range, which can influence the determined value of the circadian metric. Although the results achieved by Dai et al. [2018] are very interesting, they were based only on simulation calculations and had not been verified by measurements in real conditions. Further research in this area should be carried out, because the results could be very useful for lighting design programs.

Nevertheless, at the moment it is worth knowing that energy-efficient circadian lighting solutions can probably be achieved by:

- increasing the room-surface reflectance, but taking into account spectral reflectance or absolute circadian reflectance (determined by using the same spectral effectiveness function as the circadian metric considered),
- indirect luminaires and a highly reflective ceiling,
- optimized lighting spectra for the highest circadian efficacy.

An interesting experimental study [Hartman et al. 2016] was carried out to quantify the effect of interior surface colors (of the walls, ceiling, and floor) on circadian stimulation by daylight, where the circadian light model (*CLA*) was used. Four identical models of a standard office with the following interior colors: grey, white, blue, and yellow – were exposed to daylight (the white model was used for reference). It worth underlining that the Authors used absolute circadian reflectance for calculating the reflected component of circadian light *CLA*. It is interesting that photopic and circadian reflectance differed most for yellow surfaces (photopic: 0.65 and circadian: 0.15), while for blue (photopic: 0.45, circadian: 0.56) and grey (photopic 0.56, circadian: 0.48) surfaces, it differed moderately, and for white ones, slightly (photopic: 0.85 and circadian: 0.87) [Hartman et al. 2016]. As could be expected, the results obtained for the blue model illuminance level were lower, but circadian illuminance there reached the second highest value (after the white model). As regards the yellow model, although the greatest values of photopic illuminance were the second highest value (after the white model), circadian illuminance was the lowest, due to filtration of the blue part of the light spectrum. It shows that the incorrect choice of internal surface color can decrease circadian stimulation despite high values of photopic illuminance. For rooms with greater depth and smaller windows, the Authors recommend white or light grey surfaces, in order to obtain uniform spectral reflectance resulting in both higher illuminances and higher circadian stimulation by daylight.

9.6   POSSIBLE "TRAPS" IN LIGHTING ASSUMPTIONS

9.6.1   CCT vs Spectral Power Distribution (SPD)

Correlated color temperature (CCT) is not an accurate metric for selecting and specifying lighting to promote circadian entrainment; the reason it is still used is because it is familiar to designers.

Assigning a given light source type in a catalog, e.g. 4000 K does not mean each source of this type has exactly CCT 4000 K. According to the accepted standard [Steen, 2011], to designate the source as 4000 K, the CCT calculated from the preparatory measurements must be within the range 3985 K ±275 K (i.e. 3710–4260 K). Therefore, assuming data for six LEDs, which can be formally assigned to the 4000 K group, but which differ in spectral distribution and the actual color temperature determined for them [Aderneuer et al. 2019], the values of three circadian metrics have been determined: *CS*, a_{cv}, and *MEDI* (see Chapter 6). The actual CCT calculated from SPD and their corresponding metrics, *CS* and a_{cv}, are presented in Figure 9.2.

When we assumed that each of those sources has CCT 4000 K, we can obtain the metric ranges for this CCT presented in Figure 9.3. As it can be easily observed, the values of *CS* ranged from 0.19 to 0.46, and *MDEI* from 189 to 317 lx, which has a significant influence on designing the circadian stimulus. The less sensitive metric seems to be a_{cv}, whose range of changes is between 0.47 to 0.55.

The comparison of *EML* and *CS* metric values for different LEDs of the same CCT 3500 K, 4000 K, and 5000 K was carried out by Ashdown [2019]. Using the free, available calculators for *CS*, *M/P* ratio determination, and SPDs of different

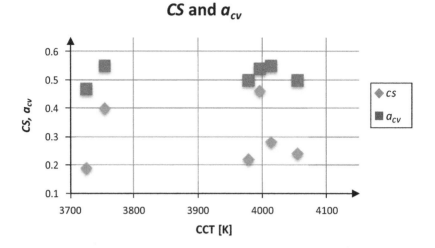

FIGURE 9.2   *CS* and a_{cv} calculated for six metameric LEDs of CCT: 3726 K, 3755 K, 3978 K, 3995 K, 4013 K, and 4054 K, which could be formally designated as 4000 K. Data from Aderneuer et al. 2019

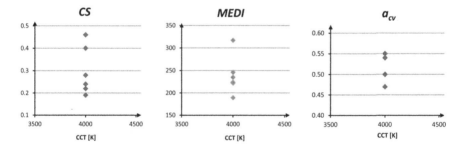

FIGURE 9.3   Variability of circadian metrics for the same assigned CCT = 4000 K of metameric LEDs: (left) *CS*, (middle) *MEDI*, (right) a_{cv}. Data from Aderneuer et al. 2019.

LEDs, he got huge ranges of vertical illuminance to obtain *CS* = 0.3. For 3500 K, the range of illuminance was from about 220 lx to 700 lx, for 4000 K from about 300 lx to 680 lx, and the smallest range of about 300 lx to 400 lx for 5000 K. The melanopic ratio, which is used for *EML* calculation (see Section 6.3) also lies in the big range for 3500 K – between 0.45 to 0.65; for 4000 K – between 0.45 and 0.75; and for 5000 K – between 0.62 and 0.75 [Ashdown, 2019].

A commonly used CCT for assessing color appearance is still used when discussing circadian lighting, because it is widely known and intuitively understood, but it only applies to the visual effect of the perceived color of the light, not its non-visual effect. However, when designing circadian lighting, one should take into account the specific circadian metric at the selected color temperature. CCT alone is not enough to determine the circadian effect of a given source.

9.6.2   CAUTIOUS EXPOSURE TO BLUE LIGHT

The use of blue or cool white light with a high proportion of blue light is recommended in integrative lighting for circadian system stimulation, but also the other side of the coin has to be considered – the potential negative effects of blue light. Therefore, when designing, care should be taken to avoid excessive (in intensity and duration) use of this type of light. This is especially the case when we use task lighting in order to increase the value of the circadian stimulus. There are three main aspects to which attention should be paid when using blue light for the circadian stimulus:

- the blue light hazard for the human retina and related photochemical injuries; exposure limit values are regulated by Directive 2006/25/CE [Directive 2006] and ICNIRP recommendations [ICNIRP 2013],
- age-related macular degeneration (AMD) caused by prolonged exposure to blue light; no exposure limit values have been developed so far,
- discomfort and blurry vision under saturated blue light exposure.

The exposure limit values (ELV) for blue light hazard are included in regulations concerning the working environment. They can be found for example in Directive 2006/25/EC [Directive 2006] or ICNIRP recommendations [ICNIRP 2013]. When choosing the light source, these ELVs must not be exceeded. It should be remembered that the photochemical effect, for example, the photochemical blue light hazard for the retina, has a summative character. Even if we choose a light source of low risk (lamp risk group built on exposure limit value of blue light hazard – effective radiance and maximum exposure time – according to EN 62471 [2006]) which is supposed not to cause hazard in normal conditions of use, we must remember that using that lamp for direct illumination of the eyes was not considered a normal condition of use. It is worth mentioning the research by Jaadane et al. [2015] concerning retinal cell damage by blue LEDs of different wavelengths and proven oxidative damage and retinal injury when using much lower radiances than those established as ELV in the current regulations (i.e. $L_B = 100$ Wm^{-2}sr^{-1} for exposure time greater than 10,000 s) given by ICNIRP and the European Directive [Directive 2006/25/EC; ICNIRP 2013]. In their conclusion, the authors stated that current regulations for the blue light hazard should be re-evaluated by transposing their results to the human eye [Jaadane et al. 2015]. These results are in agreement with the study of Behar-Cohen et al. [2011] which illustrates the curve of ELV (effective radiance of blue light hazard, L_B) and calculated effective radiances for different exposure times of six different blue LEDs and the same radiant flux equal to 0.5 W (presented in Figure 9.4). The ELVs were calculated according to the European Directive [Directive 2006/25/EC]. The maximum permissible time of exposure was between 15 s and 20 s [Behar-Cohen et al. 2011].

The other very important issue is the potential hazard for the macula on the retina, which is very sensitive to blue light. The macula is a special area in the center of the retina which is responsible for visual acuity. Neither the retina nor the macula can be regenerated or replaced if damaged. Emerging research [Schen et al. 2016] suggests that cumulative and constant exposure to blue light can damage retinal cells

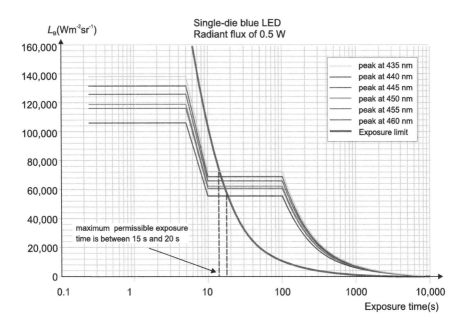

FIGURE 9.4 Graph showing the variation of L_B with the exposure time determined for six types of blue LEDs emitting a radiant flux of 0.5 W. The bold curve is the exposure limit value. Taken from Behar-Cohen 2011, with permission.

and lead to age-related macular degeneration (AMD). It is important to note that the current regulations and standards related to blue light hazard evaluation have been established on the basis of acute light exposure and do not take into account the effects of repeated exposure [Krigel et al. 2016]. This should be considered if we want to use blue-enriched light for circadian stimulus every day.

The third aspect of blue light is related to blurred vision and visual discomfort. Blue cones are spaced on the retina more in the peripheral part than in the center – with more green and red cones. There are sparse blue cones in the macula, so under saturated blue light, visual acuity is much reduced and vision is blurred, which leads to difficulty in reading as well as to other eye strain symptoms, such as glare sensitivity, tired eyes, and dry eyes.

The above-mentioned negative effects of blue light raise doubts about the correctness of using desktop lighting with blue light directed straight into the eyes, in the way it is recommended, as one of the possible variants to deliver high circadian stimulus by LRC [LRC 2019c] (see Section 8.2.3). Such solutions are certainly not recommended for people with suspected AMD or other retinal diseases that contribute to sensitizing the retina to the blue light hazard. Perhaps the future solution will be to use new violet-pump-LEDs for lighting that is safer for the retina (see Section 8.2.3).

9.6.3 CIRCADIAN LIGHTING AND GLARE

Although glare is considered one of the quality parameters of lighting, it should be noted that with lighting that provides relatively high illuminance values at the eyes, there will be a much greater likelihood of glare than before. In addition, direct glare from local (desktop) lighting luminaires should also be included, as well as reflected glare, e.g. from computer screens. Currently, it may be worth considering the use of indirect lighting. Although less energy-saving, it enables easier avoidance of glare and provides even illumination for the cornea of the eye to achieve the circadian effect [Dai et al. 2018].

The other important fact is that discomfort glare from highly saturated colors of light has shown that yellow light is perceived as less glaring that bluish light. Another study carried out in LRC [Nagare 2017] devoted to discomfort glare assessment from LEDs of two different CCT – 3000 K and 6500 K – showed that LED – glare source of 6500 K was subjectively assessed as significantly much more glaring than 3000 K.

9.7 ESTIMATED BENEFITS OF CIRCADIAN LIGHTING FOR EMPLOYERS

Although it is difficult to predict the benefits of introducing HCL for employers, the company which acted as strategic advisor for Lighting Europe developed the estimation of economic benefits on a micro- and macro-level [Kearney 2015]. The micro model covered the perspective of individual investors and the macro model the perspective of the general public in Europe in 2020. The authors [Kearney 2015] stated that the macro-level effects of HCL require further justification based on long-term studies to define economic benefits, but some micro-level effects of HCL were developed. The chosen estimated benefits for investors are presented in Table 9.1.

TABLE 9.1
Estimated Benefits of Introducing HCL

	Industry		
Benefit	**Repetitive tasks**	**Advanced manual tasks**	**Offices**
Increased productivity (due to increased alertness and energizing effect):			
• more productivity	4.5%	4.5%	1.15%
• fewer errors	1%	2%	–
Fewer accidents due to increased alertness	1%	2%	–
Fewer sick days (due to higher physical robustness)	1%	1%	1%
Other	1-year improved retention due to higher physical robustness	1-year improved retention due to higher physical robustness	+1-year duration of workers staying

Data Based on Kearney 2015.

The micro-level estimation showed that the benefits of introducing HCL are encouraging, especially for industry, where the increase of productivity was proven. Macro-level estimation of HCL effects for Europe in 2020 (assuming a realistic market) showed a 0.87 billion EUR benefit [Kearney 2015].

9.8 SUMMARY

Designing HCL requires much more knowledge than the creation of lighting designs according to accepted standard codes. Taking into account the biological and emotional impact of lighting on the user, HCL requires familiarization with a variety of characteristics (age, gender, chronotype, health condition etc.), individual needs, and preferences. Intelligent use of color accent lights could also contribute to the user's concentration and creativity. Nevertheless, the use of blue light should always take into account the potential hazard to the retina and melatonin suppression, especially during the night.

The interaction of daylight on total user exposure to light should not be overlooked. This concerns exposure both during work time, and before or after work. For this reason, the use of different personal dosimetry devices for melanopic irradiance measurements could be used in the near future. Wireless data transfers concerning previous exposure to effective circadian light from dosimeters to an intelligent lighting control system and adjusting the appropriate artificial lighting on this basis, can significantly contribute to improving the well-being and health of workers. It seems that the direction of HCL development is personalized lighting for each worker, integrated with an intelligent lighting system. On the one hand, lighting should be provided that will automatically adjust the lighting parameters to the rhythm of daylight changes and to personal biological requirements, and on the other hand, the system should allow the intervention of those workers who would prefer to adapt them to their own needs.

As the preliminary estimates of the economic benefits of using HCL are now available, it can be assumed that the demand for designing such lighting will increase. Therefore, designers will be required to quickly adapt to the clients' expectations, which demands spending time on learning and taking advantage of the recommendations arising from scientific research. It can be said that lighting designers face a great challenge, but also a heavy responsibility for the lighting they create.

10 Measurements and Assessment of Lighting Parameters and Measures of Non-Visual Effects of Light

Always keep the direction of the light source clearly in mind and keep asking yourself, "How light is it?" and "How dark is it?"

Leoni Duff

10.1 BASIC CLASSIFICATION

Assessment of lighting parameters and estimating the non-visual effects of light is not an easy task. In most cases we use objective methods based on measurement, but also subjective ones based on the assessment of respondents given in specially developed questionnaires or surveys. The former often require both specialized (and sometimes sophisticated) equipment dedicated to measuring various parameters, which are determined either directly or indirectly, as well as expert knowledge on how to correctly make measurements of a given parameter and then arrive at the appropriate values (using advanced calculation methods) that describe the visual or non-visual effects of light. Considering the above, we have attempted to classify the measures of lighting and of the non-visual effects of light into several groups, due to their common assessment features or the type of measuring apparatus used to determine them. The following classification of methods was adopted, taking into account aspects such as spatial distribution of luminance in the field of view, spectral power distribution of light, periodic fast-changing light intensity over time (temporal distribution), and psychophysiological response to light. In addition, direct and indirect measurements of specific parameters were distinguished when analyzing particular methods.

10.2 PERIODIC FAST-CHANGING LIGHT INTENSITY OVER TIME (TEMPORAL DISTRIBUTION)

10.2.1 FLICKER ASSESSMENT

Flicker and stroboscopic effects require measurements and analysis of specific light intensity parameters which are changing over time with high operating frequencies

(Chapter 5.8). Flicker assessment is of great concern in LED-based lighting, with negative consequences resulting from the use of poorly designed solid-state lighting (SSL) products ranging from headaches to, in extreme instances, epilepsy attacks [Wright 2016]. Recommendations for the evaluation and measurement of flicker were developed by CIE in 2016 in the form of their Technical Note [CIE 2016c]. In accordance with this document, temporal light artifacts (*TLAs*) are discussed – regarding all the artifacts that are associated with changes in time and ways they can affect people in visual and non-visual ways. *TLAs* cover three independent problems: flicker, stroboscopic effect, and the Phantom Array Effect – Ghosting [CIE 2016c; NEMA 77-2017]. In the IEEE standard we can find a summary of such physiological and psychological impacts [EEE 1789]. Measurement of the parameters associated with flicker in such a wide range of cases is a difficult task. The basic parameter which is always used to describe flicker is the frequency of changes – flicker frequency (*ff*). In order to improve the characterization of this phenomenon and its nuisance, various measures (indexes) have been introduced. In the same IEEE standard [EEE 1789], the basic parameters of flicker are discussed – frequency, percentage (modulation depth), and flicker index, *FI*. Unfortunately, not all of these are directly related to human perception, so they are insufficient for evaluation, although in many applications they are still used.

Two parameters according to the IEC Standard [IEC 61000-4-15] have been introduced: short term flicker perceptibility (P_{ST}) for perceptibility over 10 min, and long term flicker perceptibility (P_{LT}) for perceptibility over 2 h. In scientific publications there are also alternative measures: flicker visibility measure – periodic (*FVM*), and flicker visibility measure – transient (*TFVM* or *FVMt*) [Perz 2019].

From the practical point of view, according to [CIE 2016c], two parameters are used in the assessment of flicker and stroboscopic effect: percent flicker (*PF*) (also modulation depth or flicker modulation) – Formula (10.1) – and flicker index (*FI*) – Formula (10.2). Percent Flicker describes changes of luminance in percent. This parameter does not take into account the frequency of changes and the waveform of changes.

$$PF = \frac{(L_{\max} - L_{\min})100\%}{L_{\max} + L_{\min}} \tag{10.1}$$

where L_{max}, L_{min} are the maximum and minimum of luminance values.

$$FI = \frac{S_1}{S_1 + S_2} \tag{10.2}$$

where S_1 is the area above the average luminance level in the graph of luminance changes, and S_2 is the area below the average luminance level.

The Flicker Index takes values in the range from 0 to 1. It is determined for one period and does not take into account the frequency of changes. The value below 0.1 is recommended [CIE 2016c]. Because *PF* and *FI* take into account only the properties of one period of change, flicker assessment often includes the frequency of changes (in Hz).

10.2.2 Stroboscopic Effect Assessment

The new measure for stroboscopic effect assessment is stroboscopic visibility measure (*SVM*) [CIE 2016c] – Formula (10.3):

$$SVM = \left(\sum_{m=1}^{\infty} \left| \frac{C_m}{T_m} \right|^n \right)^{1/n}$$

(10.3)

where C_m is the amplitude of the m-th Fourier component, while T_m is the visibility threshold for sinusoidal modulations at the frequency corresponding to C_m. According to CIE [2016c], the value of n should be defined theoretically or as a result of experiments, and it is assumed that $n = 3.7$. The value of T_m can be determined using Formula (10.4).

$$T_m = \frac{1}{1 + e^{-a(f_m - b)}} + 20 e^{-f_m/10}$$

(10.4)

where f_m is the frequency of the m-th Fourier component in Hz, $a = 0.00518$, and $b = 306.6$.

A value of *SVM* equal to or greater than 1 means that the stroboscopic effect will be visible to the observer. Unlike the flicker index and flicker percent measures, *SVM* takes into account the different sensitivities of the human imaging system (eye and brain) with respect to flicker frequency.

In addition to the *SVM* parameter, in the work of Bullough et al. [2012], other parameters concerning the stroboscopic effect have been proposed.

The d parameter that determines the percent of detection of the stroboscopic effect – Formula (10.5):

$$d = \frac{(25PF + 140)100\%}{f + 25PF + 140}$$

(10.5)

This is applicable for *PF* from range (5–100%) and f from range (100–10,000Hz).

The a parameter that determines the acceptability of the stroboscopic effect – Formula (10.6):

$$a = 2 - \frac{4}{1 + f/f_b}$$

(10.6)

where f_b is the borderline frequency between acceptability and unacceptability – Formula (10.7).

$$f_b = 130 \log PF - 73$$

(10.7)

The following scale defined the acceptability of the stroboscopic effect:

$a = +2$ very acceptable

a = +1 somewhat acceptable
a = 0 neutral (neither acceptable nor unacceptable)
a = −1 somewhat unacceptable
a = −2 very unacceptable

Measuring devices consisting of a high-performance photo detection system (a photodiode with a trans-impedance amplifier) supplemented by an optical $V(\lambda)$ filter are used to measure the flicker parameters of the light source. Modern measuring techniques make it possible to build simple measuring instruments which on the basis of recorded changes of illuminance can determine all the parameters used in the evaluation of the flicker and stroboscopic effect. The review of benchtop laboratory meters and handheld meters can be found in DOE [DOE 2016; DOE 2018].

In this context, a much more important problem than measuring parameters is the interpretation of results and the determination of safe working conditions. Research described in Bullough et al. [2012] led to the development of the standard [IEEE 1789] ensuring lighting safety in industrial conditions (Figure 10.1). This is particularly important when using LEDs in designed lighting installations.

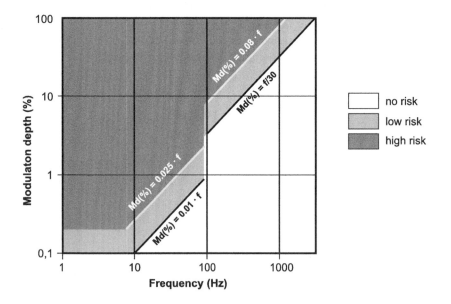

FIGURE 10.1 Practical recommendation of a combination of flicker parameters (based on IEEE 1789). Frequency in Hz and modulation depth (percentage flicker according to Formula (10.1)) in percent are taken into account. The white area means a safe combination of parameters, light gray – combinations associated with low risk, dark gray – risk. Lines delimiting these areas are described by the relationship between f (frequency) and Md (modulation depth). The equations allow determination of the Md value for the given range of f.

10.3   SPATIAL DISTRIBUTION OF LUMINANCE IN
THE FIELD OF VIEW (LUMINANCE MAP)

Assessment of the luminance distribution in the appropriate field of view, including angular relationships of glare source position and its angular dimension, requires both the appropriate measuring device, measurement procedure, and advanced calculation methods for assessment performance. The image luminance measuring device (ILMD) is currently the most advanced instrument for determining the map of the luminance distribution in a specific angle of field of view [Błaszczak 2013]. Two groups of ILMD solutions are used in practice. The first group consists of relatively expensive stationary-benchtop measuring instruments intended for laboratory tests [TechnoTeam 2019a]. These devices are sometimes also used in field measurements. The second group contains "consumer devices" that are used in any non-laboratory conditions. These are built based on digital cameras [TechnoTeam 2019b]. Correct calibration and specialized software ensure correct luminance distribution measurements. Such software is usually part of a measuring system. Independent software [Bellia et al. 2013] can also be used, for example, the EVALGLARE system [RADSITE 2019] cooperating with the RADIANCE simulation software [Radiance 2019]. These measuring devices are most often used for measuring luminance distribution and glare evaluation by determining the appropriate *UGR*, *DGP* (*DGI*), or *TI* index based on luminance distribution measurements and angular dependencies.

A major technical problem is the range of luminance that is measured. Humans are able to see a picture in a very wide range of luminance: from 0.000001 cd/m^2 to 1,000,000 cd/m^2 [Kolb 1995] – the scope of cones and rods. Currently produced CCD or CMOS sensors, even for special purposes, do not cover such a huge range of luminance.

The tonal range of the object recorded in the image can be assessed using the *EV* (exposure value) parameter. It is defined by Formula (10.8) [ISO 2721] and describes the lighting conditions for photography [Jacobson et al. 2000]. Increasing the *EV* by one degree corresponds to increasing the aperture by one degree or doubling the exposure time (by default, in cameras it is also a changed by one degree on the timescale). More information on the basic parameters affecting the tonal range and on the use of these parameters in radiometry can be found in the article by Hamey [2005].

$$EV = \log_2 \frac{f_{AV}^2}{TV} \tag{10.8}$$

where: f_{AV} – relative aperture value, TV – exposure time in second (shutter speed).

The value of the *EV* parameter is directly related to a specific sensor (CCD or CMOS) of the camera and is given for a specific sensitivity. This is usually 100 ISO, but sometimes different sensitivity is required and then the *EV* value is converted [Sawicki et al. 2013]. From the formal point of view, the *EV* parameter is a measure of neither illuminance nor luminance. However, with a properly exposed image, the *EV* value is proportional to the illuminance at a given point [Ray 2002].

Using the *EV* parameter, we can specify the tonal range that is needed to obtain a correctly exposed image. For a typical office room, if the view does not include light sources, this range is about 6–8 *EV*. The appearance of light sources causes an increase to 20 *EV* or more. It is assumed that owing to the work of cones and rods, the human visual system is able to correctly distinguish details in the tonal range of about 14 *EV*. The tonal range of a typical photographic sensor (CCD or CMOS) is only about 8–10 *EV*, taking into account noise (at a resolution of 3 × 8 bits per pixel) [Cambridge in Colour 2019]. The phenomenon of glare occurs when a light source of high luminance appears in the field of view – the tonal range in this case may exceed the standard human perception capabilities – hence the problem of indirectly arising discomfort. However, the measurement evaluation of such a phenomenon would require a sensor that could record the image correctly. No sensor produced today meets this condition.

To solve this problem, the technique of photos with increased tonal range (HDR, HDRI – high dynamic range imaging) is used [Reinhard et al. 2006]. This technique involves building a final photo (with an increased range) based on several component photos, each taken under different exposure conditions.

The first attempt to use HDRI was a photograph of the Memorial Church in Stanford made by one of the authors of the method [Debevec et al. 1997]. The final image was built from 16 component images. Such a large number of photos made it possible to cover all of the necessary tonal range. Pictures for the purposes of measuring luminance distribution and glare assessment are taken with photographic equipment that normally allows only three pictures to be taken automatically (in AEB mode) with different exposure conditions (Figure 10.2). This solution allows only the expansion of the tonal range. It can be shown [Sawicki et al. 2013] that such use of HDRI technique for measurements with the LMK photometer and thus with typical equipment in this case, offers the possibility of extending the tonal range by about 5 *EV*. Thanks to this, a range of approx. 14 *EV* is available for measuring luminance.

To effectively use this range, the exposure conditions for individual component photos should be selected very precisely. This problem is widely discussed in the literature [Bellia et al. 2013; Boulenguez et al. 2015; Grossberg et al. 2004; Hamey 2005]. The authors Sawicki et al. [2013] and Wolska et al. [2020] proposed an effective algorithm for determining exposure time. In the first version of the algorithm [Sawicki et al. 2013], selection of the correct exposure time was carried out using a feedback loop in which the luminance measurement is validated. This was complicated and time-consuming. To simplify and speed up the procedure, a few hundred measurements of *UGR* were analyzed. On this basis, a formula for determining exposure time based on the maximum luminance of glare sources has been proposed [Wolska et al. 2020]. This algorithm has repeatedly been successfully used for *UGR* and *GR* measurements at various workplaces.

It seems that the simplest solution for recording HDR images for luminance measurement and glare assessment would be to take as many photos as possible under different exposure conditions and then apply an effective image reconstruction algorithm with a wide tonal range. The review of the method and algorithms concerning HDR is presented in Mantiuk et al. [2015]. At the same time, many different formats of saving HDR image to file are used. The OpenEXR format [OpenEXR 2019] is the open standard in photography today. It uses a 32 bit floating points number per pixel. A comparison of different formats and their properties can be found in Ward [2019].

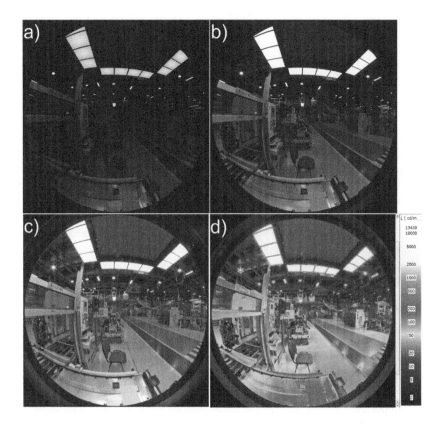

FIGURE 10.2   An example of HDRI assembly from components. Three component images: (a) –2 *EV*, (b) 0 *EV*, (c) +2 *EV*. (d) Luminance map as a result of the assembly.

Unfortunately, most known reconstruction methods were developed for the needs of photography and computer graphics. Direct application of these methods in measurement technology does not always bring the desired results. The main problems are noise and the non-linearity of the tonal processing characteristics. A kind of confirmation of these problems is the fact that software manufacturers for luminance measurements (e.g. LMK) use their own, independently developed algorithms and their own data recording formats.

10.4   GLARE ASSESSMENT

Nowadays glare is assessed in three ways:

- subjective assessment using an appropriate semantic scale,
- simulation assessment using an appropriate glare index and a specialized simulation program,
- objective (measuring) assessment using an appropriate glare index and measuring system.

Proper glare assessment is very important. Relying only on lighting design and simulation calculations does not always yield the desired results. The only way to assess glare in real conditions is to take measurements. However, all three methods are used in practice.

10.4.1 SUBJECTIVE ASSESSMENT OF GLARE

Subjective glare has always been a big challenge for researchers due to perceptual problems [Hirning et al. 2014]. Subjective assessment is carried out among workers using appropriately developed questionnaires. Participants complete a survey by answering questions about their subjective visual impressions. The survey is based on using an appropriate semantic scale to describe the workers' impressions. Originally, the purpose of such studies was to facilitate the subjective assessment of glare. Over time, they also began to be used to compare research carried out under different conditions, and thus to compare glare indexes as a measure of the phenomenon. Nowadays, de Boer and Hopkinson scales are used [Hopkinson 1957; Hopkinson 1972]. There exist many glare indexes and applications of scales [Carlucci et al. 2015]. However, from the worker's point-of-view, the feeling of being exposed to glare is more important than which mathematical models and scales are used. Therefore, research was undertaken to compare different scales. An example of unification for different semantic scales is the result of the studies described in Sawicki et al. [2016a]. The authors conducted a relationship analysis that describes glare indexes for indoor and outdoor workplaces which has proven that it is possible to unify the semantic scales for *UGR* and *GR* (Glare Index) (Table 10.1). This result is very important, because the *GR* index can also be used to assess glare at indoor workplaces [EN 12464-2].

10.4.2 SIMULATION ASSESSMENT OF GLARE

At the lighting design stage, use is commonly made of simulation assessment. Specialized lighting software such as Relux [2019] or DIALux [2019] is employed for this purpose. The analysis of light propagation in such programs is conducted by simulation methods [Pharr et al. 2016]. The most commonly used one is radiosity [Dutré et al. 2003] or, less often, ray tracing [Suffern 2007]. It is worth noting that the radiosity method is an excellent way to simulate the propagation of light in an environment where there are reflections (and/or refractions) of light in a diffused, or at least directional-diffused, form. However, the simulation of mirror reflections is a very serious unsolved problem. It is not possible to determine the luminance distribution in DIALux, when indirect lighting was designed with the use of mirror reflection. The simulation method is an excellent solution at the stage of lighting design. However, it does not allow the assessment of glare in real conditions. The studies described in Sawicki et al. [2015] lead to the conclusion that the *UGR* values determined on the basis of measurements (objective assessment) are closer to the feelings of participants than those determined in the DIALux simulation program.

TABLE 10.1

The Proposed Unified Scale for *GR* and *UGR*

GR	Glare sensation	UGR
90	Unbearable	37.5 (*)
77	Intolerable	34
70	Disturbing	32
67	Just intolerable	31
57	Uncomfortable	28
50	Just admissible	26
48	Just uncomfortable	25
39	Unacceptable	22
31	Just acceptable	19
30	Noticeable	18.5
24	Perceptible	16
17	Just perceptible	13
10	Imperceptible	10

(*) The value of the *UGR* indicator for the "unbearable" level is significantly beyond the range of discomfort glare. The value proposed here is a purely theoretical extrapolation carried out on the basis of an index change trend analysis. All values above 30 should be treated similarly. However, the results of carrying out such tests and experiments show that *UGR* values correspond to human perception [Bellia et al. 2008; Cai et al. 2013].

10.4.3 Objective Assessment of Glare at Indoor Workplaces

The use of the *UGR* index for glare assessment is well documented. ILMD makes it possible to measure the map of luminance distribution in an effective way [Błaszczak 2013]. Based on the results, specialized software determines the value of the glare index. The ILMD device built on the basis of a digital camera is most often used in practical applications.

Assessment of glare by means of evaluating the value of the glare index is a very complex process. Hence, a number of different important aspects to consider when such a measurement is being made:

1) Correct selection of the measuring point. Some requirements are: conversations with employees and with the management of the workplace, local vision and the precise planning of activities, including the selection of the right equipment and working time.
2) Correctly setting ILMD at the measurement place, so that the observation direction is in line with the worker's viewing direction [CIE 1995b; EN12464-1].

3) Choosing the proper lens of the ILMD device so that its viewing angle covers the viewing angle of the worker at the workplace [CIE 1995b; EN12464-1].

4) Determination of the maximum luminance of the glare source from the point of view of the observer's eye using a standard luminance meter. This task can be very important for modern LEDs with a luminance of a very high value.

5) Correct exposure of a series of photos taken in HDRI mode so that the luminance range (source of light and background) measured by the system corresponds to their real values (previously measured) [Sawicki et al. 2013; Wolska et al. 2020].

6) Taking into account (and if possible, minimizing) "photographic" errors resulting from the use of a camera, such as ILMD, for measuring the distribution of luminance [Cai et al. 2011].

7) Taking into account possible errors that may be the result of the not fully thought-out use of the software for determining glare. It is worth paying attention to checking the correctness of the measured luminance distribution after processing a series of photos into the HDRI form and the correct selection of the threshold luminance defining the background luminance (in the form of a "mask" of the luminance range).

8) Taking into account errors related to the indirect method of determining the value of the glare index.

10.4.4 Uncertainty of *UGR* Determination

The *UGR* index value is determined indirectly based on the distribution of luminance. All the indexes used to assess glare were defined in the form of mathematical formulas taking into account the respective lighting parameters and geometrical relationships. In Wolska et al. [2016], an analysis of uncertainties was carried out using the total differential method for formulas used to determine the values of *UGR* and *GR* indices. The formula for *UGR* (Formula (4.2)) can be written in a form depending on a single glare source, taking into account the measured parameters (Formula (10.9)).

$$UGR = 8\log_{10}\left(\frac{0.25 \cdot L_i^2 \omega_i}{L_u \cdot P_i^2}\right) \qquad (10.9)$$

$$\Delta_{UGR} \leq \left|\frac{16}{\ln(10) \cdot L_i}\right| \cdot \Delta_{Li} + \left|\frac{8}{\ln(10) \cdot \omega_i}\right| \cdot \Delta_{\omega i} + \left|\frac{-16}{\ln(10) \cdot P_i}\right| \cdot \Delta_{Pi} + \left|\frac{-8}{\ln(10) \cdot L_u}\right| \cdot \Delta_{Lu} \qquad (10.10)$$

that is,

$$\Delta_{UGR} \leq \frac{1}{\ln(10)}\left(\frac{2}{L_i} \cdot \Delta_{Li} + \frac{1}{\omega_i} \cdot \Delta_{\omega i} + \frac{2}{P_i} \cdot \Delta_{Pi} + \frac{1}{L_u} \cdot \Delta_{Lu}\right) \qquad (10.11)$$

where: Δ_{UGR} – measurement uncertainty (indirect determination of the UGR), Δ_{Li} – uncertainty of measurement of the average luminance of the i-th luminaire (cd/m^2), $\Delta\omega_i$ – uncertainty of measurement of the solid angle of the i-th luminaire $\omega\omega_i$ (sr), Δ_{Pi} – uncertainty in determining the index P_i, Δ_{Lu} – uncertainty of measurement of the average luminance of the background (cd/m^2).

The analysis leads to important conclusions regarding the impact of individual parameters on the uncertainty of determining the glare index. Usually the background luminance does not exceed 100 cd/m^2. However, it is assumed that discomfort glare may occur only at source luminance exceeding 600 cd/m^2 [CIE 1983]. In practice, most often the luminaire's luminance value is at least 1000 times higher than the background luminance (in the case of LEDs, the multiplier may be much higher). In both cases, the luminance is measured by the same instrument, which means that the values Δ_{Lai} and Δ_{Lb} are the same. Taking into account the relationship between source luminance and background luminance, it can be concluded that the uncertainty of background luminance measurement is a decisive factor. The impact of the uncertainty of measurement of the luminance source is significantly smaller – it can be practically negligible [Wolska et al. 2016; Sawicki et al. 2016b].

10.4.5 *UGR* DETERMINATION – THE PROBLEM OF THE ANGULAR SIZE OF GLARE SOURCES

The problem of the angular size of glare sources should be considered independently. This is especially important in the analysis of glare from the LEDs commonly used today [Khan et al. 2015]. The rules for determining the UGR index precisely define the angular limits of the observed source. It was assumed that this source should be in the range from 0.0003 sr to 0.1 sr. Therefore, both very small light sources and very large ones are negligible. CIE document [CIE 2002] extended the possibility of using the UGR index for small sources. This was due to the wide use of raster luminaires. The CIE publication assumed that in this case the luminance of the source in the UGR formula should be expressed by the quotient of the luminous intensity of the luminaire in a given direction and the apparent field of the luminous area. At the same time, it was assumed that the apparent luminous area of the source should not be less than 0.005 m^2, and the source should be at least 5° above the line of sight.

Research carried out in the 1990s [Flannagan 1999] showed that small light sources with 0.3° or 0.6° angular size which are visible in the field of view do not affect the correct distinguishing of details at all – and therefore do not cause glare. However, it should be taken into account that these tests were carried out using conventional light sources. In contrast, LEDs in the same geometrical conditions can affect glare due to their very high luminance. This was confirmed by later studies; an increase in the luminance of small sources causes an increase in discomfort [Rosenhahn et al. 2004]. Problems related to the size and geometry of the sources are decisive for assessing the impact of LEDs on glare. This is confirmed by different studies [Scheir et al. 2015; Tashiro et al. 2015] in which the authors examined the impact of LED matrices on the glare impression.

In Nonne et al. [2013], the authors analyzed the differences in glare using different light sources: traditional ones and LEDs in different versions. Because LEDs are characterized by a small apparent luminous area, the authors of the study used the *UGR* formula for small sources and the standard formula. Their conclusion was that the traditional formula provides an assessment which is closer to human perception.

The publication by Eble-Hankins et al. [2009] describes research on the subjective assessment of glare from sources with different non-uniform distribution of luminance, modeled by black and white stripes of different spatial frequency of source. It has been shown that discomfort rises with the increase of spatial frequency. On the other hand, however, a non-uniform stimulus is considered less uncomfortable than a uniform one. The impact of the luminous structure of luminaires with non-uniform luminance on discomfort was also found, especially the luminance of relatively dark parts of the luminaire and the average luminance of the glare source [Funke et al. 2015]. However, there was no effect of distance between LED points inside the luminaire, regardless of the direction of observation of luminance contrasts inside the luminaire [Funke et al. 2015].

The most radical conclusions were reached by the authors of the document by Cai et al. [2013]. The authors stated that all the formulas used so far are inappropriate to assess glare from LEDs. Other studies prove that glare assessment of complex light scenes (with small or large sources, complex luminance distributions) may require fundamental changes in the creation of glare models [CIE 2016b; Clear 2013]. Hence, there are also new proposals for *UGR* formulas for luminaires with white LEDs using various spatial arrangements in the luminaire. Scheir et al. [2015], Tashiro et al. [2011], and Wagdy et al. [2019] applied machine learning, while Safdar et al. [2018] used a neural response-based model. Research is also being conducted concerning relationships between discomfort glare and physiological responses (pupil diameter, blink rate, and blink amplitude etc.) [Hamedani et al. 2019; Hamedani et al. 2020; Stringham et al. 2011, Lin et al. 2015; Scheir et al. 2019].

10.5 SPECTRAL POWER DISTRIBUTION OF LIGHT (SPD)

10.5.1 SPECTRORADIOMETRIC METHOD

Precise measurement of the SPD of light could be done using an adequate spectroradiometer. There are two main groups of spectroradiometers: double monochromator spectroradiometers and diode array spectroradiometers. Most double-monochromator spectroradiometers are fixed installation systems for laboratory measurements, while diode array ones are portable devices for field use. More precise measurements could be obtained by using double monochromator instruments. Using diode array spectroradiometers, one must take into account stray light, which may sometimes significantly affect the results. Depending on the application and expected result of measurement, more extensive information provided by the spectrum, higher precision of measurement, and possibly an in-depth analysis can be required. The portable spectroradiometer seems to be more suitable for evaluating optical radiation in such situations. Spectroradiometers make it possible to measure spectral radiance or irradiance in various spectral ranges, depending on the measuring possibilities of a particular device. A built-in optical measuring system commonly covers visual radiation from approximately 380 nm to 780 nm. As a general-purpose instrument, such a device has high accuracy and built-in system applications which allow the assessment of many aspects of light.

Based on spectral irradiance measurements and using embedded (or additionally included) software, it is possible to determine many such parameters or measures as: illuminance, CCT, color coordinates, color rendering index – R_a, fidelity index R_f, circadian metrics (circadian light CLA, equivalent melanopic lux, EML etc.). If a given spectroradiometer does not have the appropriate software for calculating more complex circadian metrics (like CLA or CS), it is possible to implement the spectral distribution determined from the measurements into specially developed calculators available on websites for free. It is worth paying attention to the fact that the correctness of the assessment of circadian parameters depends on the correct location of the spectroradiometer input system – at the eye position on the vertical plane and taking into account the angles of incidence of light entering the eye, which are important in causing biological effects. In addition, the principles of determining circadian parameters are so complicated (Chapter 6 and Section 7.4) that a precise measurement procedure is needed.

Complete measurement procedures for the purposes of light assessment, in the field of the non-visual response has been introduced in the article by Lucas et al. [2014]. The authors have presented many technical and application details, thanks to which the description presented can be useful not only to scientists, but above all to engineers, lighting designers, regulatory authorities, and all those who will use the measurements in practice.

10.5.2 Radiometric/Dosimetric Method

In addition to general purpose spectroradiometers, a large group of specialized devices was built to measure lighting parameters and circadian metrics.

Although radiometers do not provide values of spectral power distribution (SPD) or irradiance (SPI) of measured radiation in their results, they can be assumed to belong to the group of devices which measure the power or irradiance of the optical radiation of a particular spectrum (although they provide an integrated value). The range of the measured spectrum depends on radiometer type and detector used.

The most common radiometric device is the luxmeter, which is equipped with a spectrally corrected detector for photopic action spectra $V(\lambda)$. It is widely used for illuminance measurements and depending on the detector construction and optics, it can measure illuminance at a given point in the area (horizontal, vertical etc.), cylindrical or semi-cylindrical and spherical or semi-spherical illuminance. The measurement of cylindrical illuminance is recommended in the standard [EN 12464-1] (measured in a vertical plane at the head of a person). It is also recommended for measurements made to assess the non-visual effects of light. For example, some dynamic lighting protocols use cylindrical illuminance values (Section 8.2). If we do not have a cylindrical detector, cylindrical illuminance can also be roughly determined as the arithmetic mean of the vertical illuminance values measured in at least the four directions of space.

The assessment of circadian lighting is made by means of various techniques. Arguelles-Prieto et al. [2019] describe the improvement in the ambulatory circadian monitoring of a small device (ACM, Kronowise). The authors developed an algorithm that makes it possible to determine light intensity and timing with three light sensors: a wide-spectrum (380–1100 nm) device, an infrared (700–1100 nm) one, and a sensor equipped with a blue filter. Using data from the Kronowise instrument, the authors analyze the full spectrum of visible light to distinguish between different wavelengths of light (range 380–590 nm and greater than 680 nm). This gives the opportunity to monitor light exposure in a more detailed way. The use of the HDR technique also yields good results in the assessment of circadian light. The authors of the study by Jung et al. [2018] developed a methodology to measure circadian light using HDR photography. They calculated equivalent melanopic lux (*EML*) using colorimetric (CIE XYZ) measurements and HDR photography for obtaining circadian luminance distribution.

For common assessment of light exposure, wearable devices are very useful. The paper by Price et al. [2017b] presents an overview of 11 available dosimeters in the form of wearable devices. All the devices described can be useful for analyzing light exposure and related effects on circadian rhythms and on sleep.

It is also worth paying attention to the blue light exposure, both from natural and artificial light. Actiwatch Spectrum (AWS device) was used in a study conducted in hospital among hospital workers who are exposed to a high dose of blue light. The authors tested the average level of light irradiance (from LED) in 24 h exposure. Under the conditions of the study, irradiance of blue light from artificial sources was much lower than from natural light in short daylight exposure [Udovičić et al. 2019].

There is a well-known problem with mismatching existing (manufactured) sensors to all requirements, including performance ones. An interesting solution is to try to improve the parameters needed for measuring melanopic irradiance for non-visual impact [Price et al. 2017a]. Its authors improved spectral performance and adapted the characteristics to melanopic response.

A good approach to exposure assessment of biological light can be measuring the total flux of light on the retina. For this purpose, a specialized instrument, such as the Retinal Exposure Detector can be applied [van Derlofske et al. 2000]. Using such devices, Aries et al. [2002] analyzed the relation between retinal illuminance, horizontal task illuminance, and vertical eye illuminance. In this way it was possible to characterize the lighting for both visual and non-visual effects in office space illuminated by daylight [Aries et al. 2002].

On the other hand, there is no generally accepted (and applied) methodology for the metrology of spectral response of the melanopsin receptor. There were two propositions described in the *Trends in Neurosciences* journal in 2014 [Lucas et al. 2014] and in 2017 in *Lighting Research and Technology* [Amundadottir et al. 2017]. A formal approach to the topic was put forward in one of the latest articles published in 2019 [Berman et al. 2019a]. The authors proposed a new approach consistent with the CIE definition of lighting units. A well-prepared review of the achievements related to circadian lighting measurements can also be found [Brennan et al. 2018; Knoop et al. 2019].

10.6 MEASUREMENT OF PSYCHOPHYSIOLOGICAL RESPONSE TO LIGHT

10.6.1 Basic Assessment Methods

There are several types of psychophysiological responses of the human body to light that can be measured objectively or subjectively in order to evaluate lighting and select appropriate values for light parameters producing particular effects. These include methods of measuring brain activity, electrical heartbeat activity, tracking and recording eye movements, measuring hormone levels in human biological material, and vigilant performance and sleepiness assessment tests. Electrical heart beat (heart rate variability – HRV) activity is usually assessed using the electrocardiography (ECG) method of monitoring and registering the electric current of a heartbeat.

ECG signal analysis and HRV are often used to estimate human psychophysiological status. A good example is the publication by Vicente et al. [2016], in which the HRV signal received from the ECG is analyzed. The authors presented research showing the possibility of detecting drowsiness using HRV analysis. Two states: "awake" and "drowsy" were compared using several features based on HRV. The HRV signal was divided into two sub-bands: low-frequency (LF 0.04–0.15 Hz) and high-frequency (HF 0.15–0.4 Hz). It is suggested that a "drowsy" state is associated with the HF sub-band and "fatigue" states are associated with increased energy in the LF sub-band. Relaxation should be associated with low levels in both the sub-bands.

Practical use of ECG analysis is very difficult due to the complex arrangement of the electrodes. Hence, attempts to replace ECG signals with other signals have been made. In the paper by Lee et al. [2017], a comparison of ECG signals and changes in blood flow (photoplethysmogram – PPG signal) for drowsiness/fatigue detection was presented. The authors also use heart rate variability to detect drowsiness, but instead of using an ECG signal, they used the PPG signal, which can be easily registered using a wristband.

Among the methods of recording eye movements, we can distinguish electro-oculography and video-oculography. Electro-oculography (EOG) is a method based on measuring the differences in the bioelectric ranges of the muscles located in the ocular area. Based on the amplitude of the computational signal, the distance between rapid changes in eye indicators (also called saccades) is determined. Movements of the eye relative to the surface electrodes placed around the eye produce an electrical signal that corresponds to eye position. Video-oculography (VOG) or eye-tracking is a non-invasive, video-based method of measuring horizontal, vertical, and torsional position components of the movements of both eyes. An *eye-tracker* can detect the presence, attention, and focus of the user.

The circadian melatonin rhythm is currently the most commonly used circadian phase marker in humans and because of that, measuring hormones and particularly melatonin levels in such human biological material as blood, urine, and saliva is one of the most popular objective methods of assessing melatonin suppression resulting from exposure to particular spectra of light. The lower the melatonin level, the higher the influence of light. As was stated before in this book (see Chapters 3, 6, 7, and 8), blue light is most effective in melatonin suppression. Research related to such measurements is carried out during the night, when the level of melatonin in the human body is at its highest. The results reveal possible nocturnal melatonin suppression by light. This method is not reliable in measuring the impact of light during the day. Besides, it is a typical medical analysis, completely impractical in industrial and office conditions.

The other commonly used method for evaluating alertness are psychomotor vigilance tests (PVT). Usually those are computer-based tests which measure an individual's reaction to specific light changes in the environment. Time, errors, and omissions are measured and recorded. On the basis of the results obtained, it is possible to draw conclusions related to alertness. The most frequent PVT tests used in studies concerning light influence on alertness are the GO NO GO and N-back tests. The standard PVT test lasts ten minutes, which appears to be an adequate duration to induce time-on-task effects in sleep-deprived individuals. However, shorter PVT variants have been successfully used to show performance decrements. The most popular on-duty alertness test is the 1–9-point Karolinska Sleepiness scale (KSS).

10.6.2 BRAIN ACTIVITY RECOGNITION: EEG AND fMRI

Two basic methods of testing brain activity can be distinguished: electroencephalography (EEG) for electrophysiological monitoring and recording electrical brain activity, and functional magnetic resonance imaging (fMRI) to detect changes related to blood flow in the brain.

The EEG technique is mostly used as a practical method to investigate the alertness state from the registered EEG signals of the human brain. EEG fluctuations apparently arise from simultaneous changes in brain mechanisms controlling central arousal and alertness and in the levels of coherent neural activity at several characteristic neural oscillation frequencies. The relationship between changes in performance and the EEG spectrum during sleepiness makes it possible to make practical use of an EEG-based real-time alertness estimation as an effect of light exposure.

The fMRI technique relies on the fact that cerebral blood flow and neuronal activation are coupled. When an area of the brain is in use, blood flow to that region increases. The fMRI concept is based on the same technology as magnetic resonance imaging (MRI). It is a non-invasive test that uses a strong magnetic field and radio waves to create detailed images of the body, but instead of looking at organs and tissues (like MRI), fMRI examines blood flow in the brain to detect areas of activity. In the area of light influence on brain activity, this method is used to observe functional brain responses during the period of exposure to a particular color of light. Based on this observation some conclusions related to the influence of light color on working memory performance, cognitive functions, and alertness can be derived. Previous knowledge suggests that blue light exposure stimulates cognitive brain activity that underlies many aspects of cognitive performance, especially the working memory. It was stated that even a relatively brief, single exposure to blue light has a subsequent beneficial effect on working memory performance [Alkozei et al. 2016].

Two techniques of brain activity monitoring are most commonly used for laboratory testing of light effects: fMRI [Alkozei et al. 2016; Vandewalle et al. 2007] and

EEG [Rahman et al. 2014; Sahin et al. 2014]. Because of the complexity of performing fMRI analysis, EEG tests are more common.

10.6.3 LIGHT IMPACT ON ALERTNESS

The effect of light on alertness has been known for many years [Figueiro et al. 2009; Figueiro et al. 2010; Figueiro et al. 2016c; Łaszewska et al. 2017; Scheuermaier et al. 2018]. Two categories of impact can be considered:

- exposure to blue light from a fairly wide wavelength range: usually from 460 nm to 480 nm. In this case, light inhibits melatonin production (Chapter 3),
- exposure to red light in the range of approximately 630 nm to approximately 640 nm. In this case, light does not affect melatonin levels, but indirectly impacts alertness in a way that has not yet been fully recognized.

The EEG test has been known for many years and for over 50 years it been used in medicine. The 10–20 electrode positioning system and its extension to a larger number of electrodes (32 or 64) is an international medical standard and makes it possible to compare the results of works by different authors. In addition to traditional medical solutions, many other non-medical applications are in use. In such cases EEG equipment does not require a conductive gel, as in traditional medical examinations. Moreover, in many cases wireless connection to a computer is possible. Such solutions significantly improve the comfort of participants. A good example is the Emotiv EPOC system [EMOTIV 2019], originally designed for commercial solutions, i.e. computer games. The ability to correctly analyze EEG signals using this equipment has been confirmed in many publications. This device is not a professional tool used for medical EEG testing; however, it is increasingly often applied in scientific research related to non-medical purposes and the analysis of EEG signals [Badcock et al. 2013; Fakhruzzaman et al. 2015]. In many publications we can find the confirmation of its correct use in non-medical applications [Ekanayake 2015; Badcock et al. 2013]. It is worth noting two problems with recording EEG signals which are important in assessing the level of alertness. The first one is the choice of EEG signals and the method used for their analysis, and the second is the choice of electrodes for recording signals.

In the literature we can find a correlation between the level of alertness and the level of the proper bands of the EEG signals. In this way it is possible to look for alertness symptoms directly in the signal analysis [Sahin et al. 2014]. The most commonly used bands for such uses are: Alpha (8–12 Hz) or AlphaTheta (5–9 Hz) [Sahin et al. 2013; Baek et al. 2015; Okamoto et al. 2014]. In addition, it is reported that the increase of alertness is associated with a decrease of the Alpha level [Lal 2001].

The problem of individual differences in EEG responses to stimuli, especially ones related to alertness, is well-known. Authors of experiments try to solve this problem by recording a very wide range of EEG signal frequencies: for example, the range that covers the sum of Alpha and Theta bands [Chang et al. 2013]. However, too wide a range can cause the averaging of details (local extremes). To solve this problem effectively, a new measure of alertness has been introduced by Sawicki et al.

[2016c], i.e. $TAAT_{max}$ (10.12) based on the capture of signal decreases in three bands: Theta, Alpha, and AlphaTheta.

$$TAAT_{max} = \max(DIFF_T, DIFF_A, DIFF_{AT})$$ (10.12)

where $DIFF_T$ is the difference of power in the Theta band. It is calculated as power_ before – power_after (before and after light exposure). $DIFF_A$ is the difference of power in the Alpha band, and $DIFF_{AT}$ is the difference of power in the AlphaTheta band, similarly calculated. Because the decrease in the signal level is correlated with an increase in alertness, the greater the $TAAT_{max}$, the higher the level of alertness.

The experiments were carried out on a large group of 50 participants using the Emotive EPOC headset. Statistical analysis confirmed the correctness of the new measure of alertness [Sawicki et al. 2016c]. The usefulness of the $TAAT_{max}$ measure has also been confirmed in later research [Wolska et al. 2018] carried out using a standard (medical) EEG device – 32 Ag/AgCl active electrodes in the 10–20 international system connected to a 256-channel g.Hlamp amplifier (Guger Technologies, Graz, Austria).

10.6.4 SIGNALS FROM WHICH ELECTRODES SHOULD BE CONSIDERED IN THE ALERTNESS ASSESSMENT

There is no unified (or standardized) way of analyzing the EEG signal for research on alertness assessment. This applies not only to methods, algorithms, and selections of EEG signal bands, but also to the selection of electrodes from which the EEG signal is taken. In almost all cases, similar equipment for EEG registration and placement of electrodes according to the International 10–20 system is used. There are many studies concerning the influence of light exposure on alertness, where completely different regions of the brain (and therefore different electrodes) are used for registering EEG signals:

- midline central [Łaszewska et al. 2017; Figueiro et al. 2009],
- motor cortex [Lavoie et al. 2003],
- frontal and occipital from midline central [Donskaya et al. 2012],
- anterior temporal lobes, parietal lobe, occipital lobe [Phipps-Nelson et al. 2009],
- motor cortex and midline central [Scheuermaier et al. 2018],
- motor cortex, parietal lobe, occipital lobe, and frontal lobe [Chellappa et al. 2017].

In many cases the selection of electrodes is simply based on the level of the signal – the electrode is selected if there is a higher signal for a particular band.

In the paper by Wolska et al. [2019], an experiment is described where blue and red light was used to stimulate alertness and the changes in the level of alpha and beta bands were analyzed. As the impact of such colors of light has been documented, alertness changes due to a specific color of light were evaluated. The experiment was

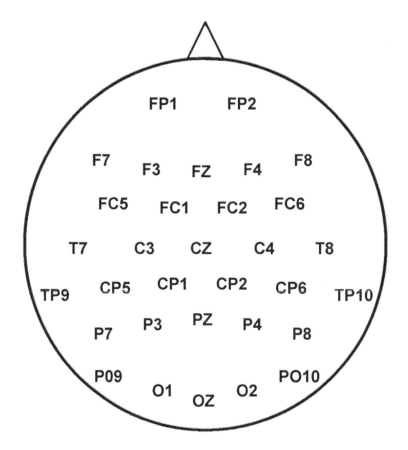

FIGURE 10.3 The layout of 32 electrodes on the scalp in the experiments conducted.

carried out on a large group of 33 participants, using a 10–20 EEG system with 32 electrodes (Figure 10.3). Statistical analysis confirmed that it is possible to select the right electrodes in such an experiment. As a result of such analysis, a set of impact maps was obtained, where the influence of light on the signal of a given electrode (in a given region) is marked by the color. Examples of such maps are presented in Figure 10.4. The brighter the field, the stronger the impact. Details of the analysis can be found in Wolska et al. [2019]. In order to assess alertness induced by blue and red light, C3 and FC1 electrodes should be considered for the Alpha band and F3 and FP1 for the Beta band.

10.7 SUMMARY

Making an objective assessment of the impact light has on humans is a difficult task. Today, in selected areas, the problem seems to have been largely solved. Glare assessment is a good example. Many years of research have led to the development of mathematically well-refined measurement methods. At the same time the development of technology has led to effective practical solutions. Nevertheless, new light

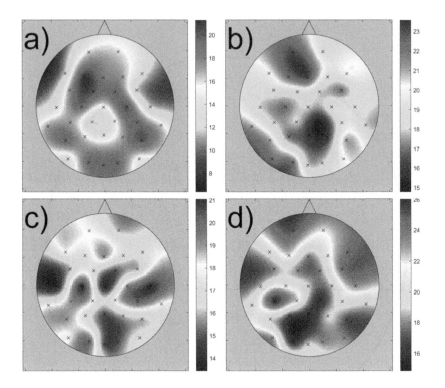

FIGURE 10.4 A set of impact maps with gray scale. The brighter the field, the stronger the impact. (a) blue light influence Alpha band; (b) blue light influence Beta band; (c) red light influence Alpha band; (d) red light influence Beta band.

sources (LEDs) pose new challenges, and research on the correct assessment of glare from these is continued. On the other hand, estimating human psychophysiological parameters and the impact of spectral properties of light on people are still very difficult tasks. Of course, it would be possible by installing many sensors on workers' bodies, using ECG and EEG and simultaneously measuring blood and urine parameters. Specialists in physiology and medicine could certainly add some interesting methods of measurement and analysis. A set of the results obtained would provide the opportunity to conduct very interesting research and develop many valuable publications. However, such tests would only be possible in hospital settings. It is not possible to apply such methods in working environments. Any attempt to apply body sensors in industrial conditions would immediately be opposed by workers. Fortunately, the development of technology and measuring techniques creates possibilities to significantly simplify this problem. The use of indirect measurements means that assessment can be made without implementing methods considered invasive. We know a lot about the influence of light on the human body, especially in the field of the non-visual impact. This results from the recent very strongly developed research in this area. Owing to this, many psychophysiological parameters related to the state of human well-being can be measured indirectly on the basis of assessing the spectral properties of light. In addition, there is a clear tendency to simplify the estimation methods within the scope discussed here. Taking into account the measurement of appropriate light distribution [Aries et al. 2002; Knoop et al. 2019], evaluation can be carried out by means of small and portable devices [Udovicic et al. 2019; Arguelles-Prieto et al. 2019]. Today it is already possible to control the spectral parameters of light at shift workplaces (Section 8.2) and simultaneously assess the parameters using wearable dosimeters [Aries et al. 2002; Knoop et al. 2019, Price et al. 2017b]. It can be assumed that if this trend persists, in a few years' time workers will be equipped with simple dosimeters; making it possible to assess lighting and indirectly also evaluate their psychophysiological parameters.

References

Aderneuer, T., O. Stefani, O. Fernández, C. Cajochen, and R. Ferrini. 2019. Circadian tuning with metameric white light: Visual and non-visual aspects. *Preprints* 2019090227. DOI: 10.20944/preprints201909.0227.v1.

Akashi, Y., R. Muramatsu, and S. Kanaya. 1996. Unified Glare Rating (UGR) and subjective appraisal of discomfort glare. *Lighting Res Technol* 28(4):199–206.

Akerstedt, T., and M. Gillberg. 1990. Subjective and objective sleepiness in the active individual. *Int J Neurosci* 52(1–2):29–37.

Alkozei, A., R. Smith, D. A. Pisner et al. 2016. Exposure to blue light increases subsequent functional activation of the prefrontal cortex during performance of a working memory task. *Sleep* 39(9):1671–1680. DOI: 10.5665/sleep.6090.

Allen, A. E., F. P. Martial, and R. J. Lucas. 2019. Form vision from melanopsin in humans. *Nat Commun* 10(1):2274.

Allen, E. A., R. Storchi, F. P. Martial, and M. A. Petersen. 2014. Melanopsin-driven light adaptation in mouse vision. *Current biology* 24(21):2481–2490.

Amundadottir, M. L., S. W. Lockley, and M. Andersen. 2017. Unified framework to evaluate non-visual spectral effectiveness of light for human health. *Lighting Res Technol* 49(6):673–696. DOI: 10.1177/1477153516655844.

Andlauer, P., A. Reinberg, L. Fourre, W. Battle, and G. Duverneuil. 1979. Amplitude of the oral temperature circadian rhythm and the tolerance of shift-work. *J Physiol* (Paris) 75:507–512.

Andresen, M., J. Mardaljevic, and S. W. Lockley. 2012. A framework for predicting the non-visual effects of daylight: Part 1: Photobiology-based model. *Light Res Technol* 44(1):37–53. DOI: 10.1177/1477153511435961.

ANSI/IES TM-30–15. 2015. IES method for evaluating light source color rendition.

ANSI/IES TM-30–18. 2018. IES method for evaluating light source color rendition.

Arendt, J. 2010. Shift work: Coping with the biological clock. *Occup Med* 60(1):10–20. DOI: 10.1093/occmed/kqp162.

Arguelles-Prieto, R., M.-A. Bonmati-Carrion, M. A. Rol, and J. A. Madrid. 2019. Determining Light intensity, timing and type of visible and circadian light from an ambulatory circadian monitoring device. *Front Physiol* 10:822. DOI: 10.3389/fphys.2019.00822.

Aries, M. B. C., S. H. A. Begemann, L. Zonneveldt, and A. D. Tenner. 2002. Retinal illuminance from vertical daylight openings in office spaces. In: *Proceedings of Right Light Conference*, May 2002, Nice, France, 75–80.

Ashdown, I. 2019. Circadian lighting: An engineer's perspective. IES. https://www.ies.org/fires/circadian-lighting-an-engineers-perspective/ (accessed October 17, 2019).

Athanasiou, D., M. Aguila, J. Bellingham et al. 2018. The molecular and cellular basis of rhodopsin retinitis pigmentosa reveals potential strategies for therapy. *Prog Retin Eye Res* 62:1–23. DOI: 10.1016/j.preteyeres.2017.10.002.

Badcock, N. A., P. Mousikou, Y. Mahajan, P. de Lissa, J. Thie, and G. McArthur. 2013. Validation of the Emotiv EPOC® EEG gaming system for measuring research quality auditory ERPs. *Peer J* 1:e38. DOI: 10.7717/peerj.38.

Baek, H., and B. K. Min. 2015. Blue light aids in coping with the post-lunch dip: An EEG study. *Ergonomics* 58(5):803–810. DOI: 10.1080/00140139.2014.983300.

Bakshi, A., and K. Ghosh. 2017. A neural model of attention and feedback for computing perceived brightness in vision. In: *Handbook of neural computation*. Chapter 26. London, San Diego, CA: Academic Press, 487–513. DOI: 10.1016/B978-0-12-811318-9.00026-0.

Bargary, G., M. Furlen, P. J. Raynham, J. L. Barbur, and A. T. Smith. 2015a. Cortical hyper-excitability and sensitivity to discomfort glare. *Neuropsychologia* 69:194–200. DOI: 10.1016/j.neuropsychologia.2015.02.006.

Bargary, G., Y. Jia, and J. L. Barbur. 2015b. Mechanisms for discomfort glare in central vision. *Invest Ophthalmol Vis Sci* 56(1): 464–471. DOI: 10.1167/iovs.14-15707.

Bedrosian, T. A., and R. J. Nelson. 2017. Timing of light exposure affects mood and brain circuits. *Transl psychiatr* 7(1):e1017. DOI: 10.1038/tp.2016.262.

Behar-Cohen, F., C. Martinsons, F. Viénot et al. 2011. Light-emitting diodes (LED) for domestic lighting: Any risks for the eye? *Prog Retin Eye Res* 30:239–257. DOI: 10.1016/j.preteyeres.2011.04.002.

Behnen, P., A. Felline, A. Comitato et al. 2018. A small chaperone improves folding and routing of rhodopsin mutants linked to inherited blindness. *iScience* 4:1–19. DOI: 10.1016/j.isci.2018.05.001.

Belenky, M. A., C. A. Smeraski, I. Provencio, P. J. Sollars, and G. E. Pickard. 2003. Melanopsin retinal ganglion cells receive bipolar and amacrine cell synapses. *J Comp Neurol* 460(3):380–393.

Bellia, L., A. Pedace, and F. Fragliasso. 2017. Indoor lighting quality: Effects of different wall colours. *Lighting Res Technol* 49(1):33–48. DOI: 10.1177/1477153515594654.

Bellia, L., E. Cesarano, G. Iuliano, and G. Spada. 2008. Daylight glare: A review of discomfort indexes. *Visual Quality and Energy Efficiency in Indoor Lighting*. https://vbn.aau.dk/en/activities/visual-quality-and-energy-efficiency-in-indoor-lighting-today-for-2 (accessed October 17, 2019).

Bellia, L., F. Bisegna, and G. Spada. 2011. Lighting in indoor environments: Visual and non-visual effects of light sources with different spectral power distributions. *Build Environ* 46: 1984–1992. DOI: 10.1016/j.buildenv.2011.04.007.

Bellia, L., G. Spadaa, and A. Pedace. 2013. Lit environments quality: A software for the analysis of luminance maps obtained with the HDR imaging technique. *Energy Build* 67:143–152. DOI: 10.1016/j.enbuild.2013.08.007.

Berman, S. M., and R. D. Clear 2019b. Clear: Simplifying melanopsin metrology: Illuminating engineering society. https://www.ies.org/fires/melanopic-green-the-other-side-of-blue/ (accessed January 28, 2020).

Berman, S. M., and R. D. Clear. 2019a. A practical metric for melanopic metrology. *Lighting Res Technol* 51(8):1178–1191. DOI: 10.1177/1477153518824147.

Bernard, M., and M. G. Figueiro. 2019. Does circadian lighting work? *The Journal of the American Institute of Architects*. https://www.architectmagazine.com/technology/lighting/does-circadian-lighting-work_o (accessed January 28, 2020).

Bernardin, F., T. Schwitzer, K. Angioi-Duprez et al. 2019. Retinal ganglion cells dysfunctions in schizophrenia patients with or without visual hallucinations. *Schizophr Res* 0920-9964(19):30270-1. DOI: 10.1016/j.schres.2019.07.007.

Berson, D. M., F. A. Dunn, and M. Takao. 2002. Phototransduction by retinal ganglion cells that set the circadian clock. *Science* 295(5557):1070–1073.

Blackwell, H. R. 1981. Description of a comprehensive family of relative contrast sensitivity (RCS) functions of luminance to reflect differences in size of task detail, task eccentricity and observer age. *J Illum Eng Soc* 11(1):52–63. DOI: 10.1080/00994480.1981.10747913.

Błaszczak, U. J. 2013. Method for evaluating discomfort glare based on the analysis of a digital image of an illuminated interior. *Metrol Meas Syst* 20(4):623–634.

Bloch, E., Luo Y., and L. da Cruz. 2019. Advances in retinal prosthesis systems. *Ther Adv Ophthalmol* 11:1–16. DOI: 10.1177/2515841418817501.

Borbely, A. A., S. Daan, A. Wirz-Justice, and T. Deboer. 2016. The two-process model of sleep regulation: A reappraisal. *J Sleep Res* 25(2):131–143. DOI: 10.1111/jsr.12371.

Boudreau, P., G. A. Dumont, and D. B. Boivin. 2013. Circadian adaptation to night shift work influences sleep, performance, mood and the autonomic modulation of the heart. *PLoS One* 8(7):e70813. DOI: 10.1371/journal.pone.0070813.

Boulenguez, P. F. Gaudaire, C. Martinsons, N. Noé, S. Carré, and L. Simonot. 2015. Imaging radiometry - a fast and robust shutter speed search algorithm. In: *Proceedings of the 28th Session of the CIE*, June 28 - July 4, 2015, Manchester UK, Vol. 1, Part 2, 1385–1388.

Bovik, A. C., ed. 2009. *The essential guide to image processing*. 2nd ed. San Diego, CA: Academic Press.

Boyce, P. R. 1996. Illuminance selection based on visual performance and other fairy stories. *J Illum Eng Soc* 25(2):41–49. DOI: 10.1080/00994480.1996.10748146.

Boyce, P. R. 2014. *Human factors in lighting*. 3rd ed. Boca Raton, FL: CRC Press.

Brainard, G. C., J. P. Hanifin, J. M. Greeson et al. 2001. Action spectrum for melatonin regulation in humans: Evidence for a novel circadian photoreceptor. *J Neurosci* 21:6405–6412. DOI: 10.1177/0748730405278951.

Brancaccio, A. M., R. Enoki, C. N. Mazuski, J. Jones, J. A. Evans, and A. Azzi. 2014. Network-mediated encoding of circadian time: The suprachiasmatic nucleus (SCN) from genes to neurons to circuits, and back. *J Neurosci* 34(46):15192–15199. DOI: 10.1523/JNEUROSCI.3233-14.2014.

Brennan, M. T., and A. R. Collins. 2018. Outcome-based design for circadian lighting: An integrated approach to simulation and metrics. In *Proceeding of Building Performance Analysis Conference and SimBuild co-organized by ASHRAE and IBPSA-USA*, September 26–28, 2018, Chicago, IL, 141–148.

Brown, T. M., C. Gias, M. Hatori et al. 2010. Melanopsin contributions to irradiance coding in the thalamo-cortical visual system. *PLoS Biol* 8(12):e1000558. DOI: 10.1371/journal.pbio.1000558.

Bubl, E., E. Kern, D. Ebert et al. 2010. Seeing grey when feeling blue? Depression can be measured in the eye of the disease. *Biol Psychiatry* 68(2):205–208. DOI: 10.1016/j.biopsych.2010.02.009.

Bubl, W., M. Dorr, A. Riedel et al. 2015. Elevated background noise in adult attention deficit hyperactivity disorder is associated with inattention. *PLoS One* 10(2): e01 18271. DOI: 10.1371/journal.pone.0118271.

Bullough, J. D. 2009. Spectral sensitivity for extrafoveal discomfort glare. *J Mod Opt* 56(13):1518–1522. DOI: 10.1080/09500340903045710.

Bullough, J. D., K. Sweater Hickcox, T. R. Klein, A. Lok, and N. Narendran. 2012. Detection and hacceptability of stroboscopic effects from flicker. *Lighting Res Technol* 44(4):477–483. DOI: 10.1177/1477153511414838.

Busatto, N., T. Dalla Mora, F. Peroni, and P. Romagnoni. 2018. Circadian lighting experiences in an health residence for non self-sufficient elderly people. In: *Proceedings of BSO 2018:4th Building Simulation and Optimization Conference*, 11–12 September 2018, Cambridge, UK, 9–16. https://www.researchgate.net/publication/327837006_Circadian_Lighting_Experiences_in_an_Health_Residence_for_Non_Self-Sufficient_Elderly_People (accessed November 09, 2019).

Cai, H., and T. M. Chung. 2011. Improving the quality of high dynamic range images. *Lighting Res Technol* 43(1):87–102. DOI: 1177/1477153510371356.

Cai, H., and T. Chung. 2013. Evaluating discomfort glare from non-uniform electric light sources. *Lighting Res Technol* 45(3):267–294. DOI: 10.1177/1477153512453274.

Cajochen, C., J. Zeitzer, C. Czeisler, and D. Dijk. 2000. Dose-response relationship for light intensity and ocular and electroencephalographic correlates of human alertness. *Behav Brain Res* 115(1):75–83.

Cajochen, C., M. Munch, S. Kobialka, K. Krauchi, R. Steiner, and A. Wirz-Justice. 2005. High sensitivity of human melatonin, alertness, thermoregulation and heart rate to short wavelength light. *J Clin Endocrinol Metab* 90:1311–1316. DOI: 10.1210/jc.2004-0957.

Cambridge in Colour. 2019. Dynamic range in digital photography. http://www.cambridgeincolour.com/tutorials/dynamic-range.htm/ (accessed December 19, 2019).

Campbell, F. W., and J. J. Kulikowski. 1966. Orientational selectivity of the human visual system. *J Physiol* 187(2):437–335. DOI: 10.1113/jphysiol.1966.sp008101.

Canazei, M, W. Pohl, H. R. Bliem, and E. M. Weiss. 2016. Acute effects of different light spectra on simulated night-shift work without circadian alignment. *Chronobiol Int* 34(3):303–317. DOI: 10.1080/07420528.2016.1222414.

Cao, D., A. Chang, and S. Gai. 2018. Evidence for an impact of melanopsin activation on unique white perception. *J Opt Soc Am A-Opt Image Sci* 35(4):B287-b91. DOI: 10.1364/JOSAA.35.00B287.

Cao, Y., I. Sarria, K. E. Fehlhaber et al. 2015. Mechanism for selective synaptic wiring of rod photoreceptors into the retinal circuitry and its role in vision. *Neuron* 87(6):1248–1260. DOI: 10.1016/j.neuron.2015.09.002.

Carlucci, S., F. Causone, F. de Rosa, and L. Pagliano. 2015. A review of indices for assessing visual comfort with a view to their use in optimization processes to support building integrated design. *Renew Sust Energ Rev* 47:1016–1033. DOI: 10.1016/j.rser.2015.03.062.

Caruso, C. C. 2014. Negative impacts of shiftwork and long work hours. *Rehabil Nurs* 39(1):16–25. DOI: 10.1002/rnj.107.

Chang, A. M., F. A. Scheer, C. A. Czeisler, and D. Aeschbach. 2013. Direct effects of light on alertness, vigilance, and the waking electroencephalogram in humans depend on prior light history. *Sleep* 36(8):1239–1246. DOI: 10.5665/sleep.2894.

Chapot, C. A., T. Euler, and T. Schubert. 2017. How do horizontal cells 'talk' to cone photoreceptors? Different levels of complexity at the cone-horizontal cell synapse. *J Physiol* 595(16):5495–5506. DOI: 10.1113/JP274177.

Chellappa, S.L., R. Steiner, P. Blattner, P. Oelhafen, and C. Cajochen. 2017. Sex differences in light sensitivity impact on brightness perception, vigilant attention and sleep in humans? *Scientific Reports* 7(14215):1–9.

Chen, J. 2004. Effects of headlamp glare exposure on glare recovery and discomfort. M.S. thesis. Troy: Rensselaer Polytechnic Institute.

Cheng, P., and C. L. Drake. 2018. Psychological impact of shift work. *Curr Sleep Med Rep* 4(2):104–109.

Cheng, P., and C. Drake. 2019. Shift work disorder. *Neurol Clin* 37(3):563–577. DOI: 10.1016/j.ncl.2019.03.003.

Chraibi, S., L. Crommentuijn, E. van Loenen, and A. Rosemann. 2017. Influence of wall luminance and uniformity on preferred task illuminance. *Build Environ* 117:24–35.

CIE [Comission Internationale d'Eclairage]. 1981a. *CIE 019.21-1981: An analytic model for describing the influence of lighting parameters upon visual performance.* 2nd ed. vol.1: Technical foundations. Vienna: CIE.

CIE [Comission Internationale d'Eclairage]. 1981b. *CIE 019.22-1981: An analytic model for describing the influence of lighting parameters upon visual performance,* 2nd ed. vol 2.: Summary and application guidelines. Vienna: CIE.

CIE [Comission Internationale d'Eclairage]. 1983. *CIE 055–1983: Discomfort glare in the interior working environment.* Vienna: CIE.

CIE [Comission Internationale d'Eclairage]. 1995a. *CIE 013.3-1995: Method of measuring and specifying colour rendering properties of light sources.* Vienna: CIE.

CIE [Comission Internationale d'Eclairage]. 1995b. *CIE 117 –1995: Discomfort glare in interior lighting.* Vienna: CIE.

CIE [Comission Internationale d'Eclairage]. 2001. *CIE S 008/E. Lighting for workplaces. Part 1. Indoor.* Vienna: CIE.

CIE [Comission Internationale d'Eclairage]. 2002. *CIE 146/147:2002: CIE collection on glare 2002.* Vienna: CIE

CIE [Comission Internationale d'Eclairage]. 2014a. *CIE 212:2014: Guidance towards best practice in psychophysical procedures used when measuring relative spatial brightness.* Vienna: CIE.

CIE [Comission Internationale d'Eclairage]. 2014b. *CIE TN 002:2014. Relating photochemical and photobiological quantities to photometric quantities.* Vienna: CIE.

CIE [Comission Internationale d'Eclairage]. 2016a. *CIE S 017. ILV: International lighting vocabulary*. Vienna: CIE.

CIE [Comission Internationale d'Eclairage]. 2016b. *CIE 218:2016: Research roadmap for healthful interior lighting applications*. Vienna: CIE.

CIE [Comission Internationale d'Eclairage]. 2016c. *CIE TN 006:2016: Visual aspects of time-modulated lighting systems – definitions and measurement models*. Vienna: CIE.

CIE [Comission Internationale d'Eclairage]. 2017. *CIE 224:2017: CIE 2017 Color fidelity index for accurate scientific use*. Vienna: CIE.

CIE [Comission Internationale d'EclairageCIE]. 2018. *CIE S 026/E: 2018: CIE system for metrology of optical radiation for ipRGC-influenced responses to light*. Vienna: CIE.

CIEAASM [American Academy of Sleep Medicine]. 2014. *International classification of sleep disorders [ICSD-3]*. Darien: AASM. https://j2vjt3dnbra3ps7ll1clb4q2-wpengin e.netdna-ssl.com/wp-content/uploads/2019/05/ICSD3-TOC.pdf (accessed January 20, 2020).

Circadian ZircLight. 2019a. https://www.circadianlight.com (accessed October 17, 2019).

Circadian ZircLight. 2019b. The lighting solution for control rooms and 24/7 mission critical environments optimizes performance, productivity and health. https://circadianlight.co m/images/pdfs/Catalogs/CL_ControlRoomBro.pdf (accessed October 17, 2019).

Clark, E., and N. Lesniak. 2017. An ongoing survey of designers reveals the challenges they face and how they can be overcome. *Metropolis*. https://www.metropolismag.com/desig n/circadian-lighting-survey/ (accessed October 17, 2019).

Clear, R. D. 2013. Discomfort glare: What do we actually know? *Lighting Res Technol* 45(2):141–158. DOI: 10.1177/1477153512444527.

Costa, G. 2010. Shift work and health: Current problems and preventive actions. *Saf Health Work* 1(2):112–123. DOI: 10.5491/SHAW.2010.1.2.112.

CR EC [Commission Regulation (EC)]. 2009. Commission Regulation 245/2009 implementing Directive 2005/32/EC of the European Parliament and of the Council with regard to ecodesign requirements for fluorescent lamps without integrated ballast, for high intensity discharge lamps, and for ballasts and luminaires able to operate such lamps, and repealing Directive 2000/55/EC of the European Parliament and of the Council. *OJ L* 76:17–44.

CR EC [Commission Regulation (EC)]. 2010. Commision Regulation 347/2010 amending Commission Regulation (EC) No 245/2009 as regards the ecodesign requirements for fluorescent lamps without integrated ballast, for high intensity discharge lamps, and for ballasts and luminaires able to operate such lamps. *OJ L* 104:20–28.

CS calculator. 2019. http://www.lrc.rpi.edu/programs/lightHealth/index.asp. (Accessed, December 20, 2019).

Cuttle, C. 2017. A revised Kruithof graph based on empirical data. *Leukos* 13(1):19–20. DOI: 10.1080/15502724.2016.1159137.

Dacey, D. M., H. W. Liao, B. B Peterson et al. 2005. Melanopsin-expressing ganglion cells in primate retina signal colour and irradiance and project to the LGN. *Nature* 433(7027):749–754. DOI: 10.1038/nature03387.

Dai, Q., Y. Huang, L. Hao, Y. Lin, and K. Chen. 2018. Spatial and spectral illumination design for energy-efficient circadian lighting. *Build Environ* 146:216–225. DOI: 10.1016/j. buildenv.2018.10.004.

de Bakker, C., M. Aarts, H. Kort, E. van Loenen, and A. Rosemann. 2019. Preferred luminance distributions in open-plan offices in relation to time of day and subjective alertness. *Leukos*. DOI: 10.1080/15502724.2019.1587619.

de Boer, J. B., and D. Fischer. 1978. *Interior lighting*. Deventer, Antwergen: Kluwer Technische Boeken.

de Kort, Y. A. W., and K. C. H. J. Smolders. 2010. Effects of dynamic lighting on office workers: First results of a field study with monthly alternating settings. *Lighting Res Technol* 42:345–360. DOI: 10.1177/1477153510378150.

de Zeeuw, J., A. Papakonstantinou, C. Nowozin et al. 2019. Living in biological darkness: Objective sleepiness and the pupillary light responses are affected by different metameric lighting conditions during daytime. *J Biol Rhythms* 34(4):410–431. DOI: 10.1177/0748730419847845.

Debevec, P., and J. Malik. 1997. Recovering high dynamic range radiance maps from photographs. In: *Proceedings of SIGGRAPH'97*, August 1997. http://www.pauldebevec.com/Research/HDR/ (accessed October 17, 2019).

Demmin, D. L., Q. Davis, M. Roche, and S. M. Silverstein. 2018. Electroretinographic anomalies in schizophrenia. *J Abnorm Psychol* 127(4):417–428. DOI: 10.1037/abn0000347.

DIALux. 2019. Lighting design software. https://www.dial.de/en/software/dialux/ (accessed October 10, 2019).

Diekman, C. O., M. D. Belle, R. P. Irwin, C. N. Allen, H. D. Piggins, and D. B. Forger. 2013. Causes and consequences of hyperexcitation in central clock neurons. *PLoS Comput Biol* 9(8):e1003196. DOI: 10.1371/journal.pcbi.1003196.

DIN SPEC 5031–100. 2015. Optical radiation physics and illuminating engineering - Part 100: Melanopic effects of ocular light on human beings - Quantities, symbols and action spectra.

DIN SPEC 67600. 2013. Biologically effective illumination - Design guidelines.

Directive 2006/25/EC. 2006. Directive of the European Parliament and of the Council of 5 April 2006 on the minimum health and safety requirements regarding the exposure of the workers to risks arising from physical agents (artificial optical radiation). *OJ EC* L 114/38.

Do, M. T., S. H. Kang, T. Xue et al. 2009. Photon capture and signalling by melanopsin retinal ganglion cells. *Nature* 457(7227):281–287. DOI: 10.1038/nature07682.

Do, M. T., and K. W. Yau. 2010. Intrinsically photosensitive retinal ganglion cells. *Physiol Rev* 90(4):1547–1581. DOI: 10.1152/physrev.00013.2010.

DOE [U.S. Department of Energy]. 2016. Characterizing photometric flicker. https://www.energy.gov/sites/prod/files/2016/03/f30/characterizing-photometric-flicker.pdf (accessed January 25, 2020).

DOE [U.S. Department of Energy]. 2018. Characterizing photometric flicker. Handheld meters. https://www.energy.gov/sites/prod/files/2019/01/f58/characterizing-photometric-flicker_nov2018.pdf (accessed January 25, 2020).

Dolgonos, S., H. Ayyala, and C. Evinger. 2011. Light-induced trigeminal sensitization without central visual pathways: Another mechanism for photophobia. *Invest Ophthalmol Vis Sci* 52(11):7852–7858. DOI: 10.1167/iovs.11-7604.

Donners, M. A. H., M. C. J. M. Vissenberg, L. M. Geerdinck, J. H. F. van Den Broek-Cools, and A. Buddemeijer-Lock. 2015. A psychophysical model of discomfort glare in both outdoor and indoor applications. In: *Proceeding of 27th CIE Session*, Manchester, 602–611.

Donskaya, O. G., E. G. Verevkin, and A. A. Putilov. 2012. The first and second principal components of the EEG spectrum as the correlates of sleepiness. *Somnologie* 16:69–79. DOI: 10.1007/s11818-012-0561-1.

D'Orazi, F. D., S. C. Suzuki, and R. O. Wong. 2014. Neuronal remodeling in retinal circuit assembly, disassembly, and reassembly. *Trends Neurosci* 37(10):594–603. DOI: 10.1016/j.tins.2014.07.009.

Dumitrescu, O. N., F. G. Pucci, K. Y. Wong, and D. M. Berson. 2009. Ectopic retinal ON bipolar cell synapses in the OFF inner plexiform layer: Contacts with dopaminergic amacrine cells and melanopsin ganglion cells. *J Comp Neurol* 517(2):226–244. DOI: 10.1002/cne.22158.

Dutré, P., P. Bekaert, and K. Bala. 2003. *Advanced global illumination*. Wellesley, MA: AK Peters Natick.

Eble-Hankins, M. L., and C. E. Waters, 2009. Subjective impressions of discomfort glare from sources of non-uniform luminance. *Leukos* 6(1):51–77. DOI: 10.1582/LEUKOS.2009.06.01003.

Ecker, J. L., O. N. Dumitrescu, K. Y. Wong et al. 2010. Melanopsin-expressing retinal gan-glion-cell photoreceptors: Cellular diversity and role in pattern vision. *Neuron* 67(1):49–60. DOI: 10.1016/j.neuron.2010.05.023.

EDI 65 calculator. 2019. https://www.nsvv.nl/wp-content/uploads/2019/03/20190319-CIE-S -026-EDI-Toolbox-vE1.051.xlsx.

Ekanayake, H. 2015. P300 and Emotiv EPOC: Does Emotiv EPOC capture real EEG? http:// neurofeedback.visaduma.info/EmotivResearch.pdf (accessed January 25, 2020).

Eklund, N. H, P. R. Boyce, and S. N. Simpson. 2001. Lighting and sustained perfor-mance: Modeling data-entry task performance. *J Illum Eng Soc* 30:2:126–141, DOI: 10.1080/00994480.2001.10748358.

EMOTIV [Emotiv Systems Inc.]. 2019. https://www.emotiv.com/epoc/ (Accessed December 19, 2019).

EN 12464–1: 2011. Light and lighting. Lighting of work places. Part 1. Indoor Work Places.

EN 12464–2: 2014. Light and lighting. Lighting of work places. Part 2. Outdoor work places.

EN 13201: 2014. Road lighting.

EN 62471: 2008. Photobiological safety of lamps and lamp systems.

EN ISO 11664-1. 2011. Colorimetry. Part 1: CIE standard colorimetric observers.

Enezi, J., V. Revell, T. Brown, J. Wynne, L. Schlangen, and R. A. Lucas. 2011. "Melanopic" Spectral efficiency function predicts the sensitivity of melanopsin photoreceptors to polychromatic lights. *J Biol Rhythms* 26:314. DOI: 10.1177/0748730411409719.

Espiritu, R. C., D. F. Kripke, S., Ancoli-Israel et al. 1994. Low illumination experienced by San Diego adults: Association with atypical depressive symptoms. *Biol Psychiatry* 35(6):403–407. DOI: 10.1016/0006-3223(94)90007-8.

Esposito, A. G., L. Baker-Ward, and S. Mueller. 2013. Interference suppression vs. response inhibition: An explanation for the absence of a bilingual advantage in preschoolers' stroop task performance. *Cogn Dev* 28(4):354–363. DOI: 10.1016/j.cogdev.2013.09.002.

Ewing, P. H., J. Haymaker, and E. A. Edelstein. 2017. Simulating circadian light: Multi-dimensional illuminance analysis. In: *Proceedings of the 15th IBPSA Conference*, August 7-9, 2017, San Francisco, CA, 2363–2371.

Fakhruzzaman, M. N., E. Riksakomara, and H. Suryotrisongko. 2015. EEG wave identifi-cation in human brain with Emotiv EPOC for motor imagery. *Procedia Comput Sci* 72:269–276. DOI: 10.1016/j.procs.2015.12.140.

Feigenspan, A., and N. Babai. 2015. Functional properties of spontaneous excitatory currents and encoding of light/dark transitions in horizontal cells of the mouse retina. *Eur J Neurosci* 42(9):2615–2632. DOI: 10.1111/ejn.13016.

Figueiro, M. G., J. D. Bullough, R. H. Parsons, and M. S. Rea. 2004. Preliminary evidence for spectral opponency in the suppression of melatonin by light in humans. *Neuroreport* 5:313–316. DOI: 10.1097/00001756-200402090-00020.

Figueiro, M. G., A. Bierman, B. Plitnick, and M. S. Rea. 2009. Preliminary evidence that both blue and red light can induce alertness at night. *BMC Neurosci* 10:105. DOI: 10.1186/1471-2202-10-105.

Figueiro, M. G., and M. S. Rea. 2010. The effects of red and blue light on circadian variations in cortisol, alpha amylase and melatonin. *Int J Endocrinol* Article ID 829351. DOI: 10.1155/2010/829351.

Figueiro, M. G. 2013. An overview of the effects of light on human circadian rhythms impli-cations for new light sources and lighting system design. *J Light Vis Env* 37(2–3): 51–61.

Figueiro, M. G., and C. M. Hunter. 2016a. Understanding rotating shift workers' health risks. *OSH Occupational Health & Safety*. https://ohsonline.com/articles/2016/11/01/understa nding-rotating-shift-workers-health-risks.aspx (accessed January 25, 2020).

Figueiro, M. G., K. Gonzales, and D. Pedler. 2016b. Designing with circadian stimulus. *LD+A* 46(10):30–34.

Figueiro, M. G., L. Sahin, B. Wood, and B. Plitnick. 2016c. Light at night and measures of alertness and performance: Implications for shift workers. *Biol Res Nurs* 18(1):90–100. DOI: 10.1177/10998004155728 73.

Figueiro, M. G., B. Steverson, J. Heerwagen et al. 2017a. The impact of daytime light exposures on sleep and mood in office workers. *Sleep Health* 3:204–215. DOI: 10.1016/j.sleh.2017.03.005.

Figueiro, M. G., B. S. Steverson, J. Heerwagen, and M. S. Rea. 2017b. Circadian-effective light and its impact on alertness in office workers: A field study. In: *Proceedings of IES conference, 2017.* https://www.lrc.rpi.edu/programs/lightHealth/pdf/Figueiro_IESConference_Aug2017.pdf (accessed January 25, 2020).

Figueiro, M. G., R. Nagare, and L. L. Price. 2018. Non-visual effects of light: How to use light to promote circadian entrainment and elicit alertness. *Light Res Technol* 50(1):38–62. DOI: 10.1177/1477153517721598.

Figueiro, M. G., M. Klasher, B. S. Steverson, J. Heerwagen, K. Kampschroer, and M. S. Rea. 2019a. Circadian-effective light and its impact on alertness in office workers, *Lighting Res Technol* 51(2):171–183. DOI: 10.1177/1477153517750006.

Figueiro, M. G., L. Sahin, C. Roohan, M. Kalsher, and M. S. Rea. 2019b. Effects of red light on sleep inertia. *Nat Sci Sleep* 11:45–57. DOI: 10.2147/NSS.S195563.

Finn J. T., D. Krautwurst, J. E. Schroeder, T. Y. Cen, R. R. Ed, and K. W. Ya. 1998. Functional co-assembly among subunits of cyclic-nuclotide-activated, nonselective cation channels, and across species from nematode to human. *Biophys. J.* 4(3):1333–1345.

Flannagan, M. J. 1999. Subjective and objective aspects of headlamp glare: Effects of size and spectral power distribution. Report No. UMTRI-99-36. https://deepblue.lib.umich.edu/bitstream/handle/2027.42/49407/UMTRI-99-36.pdf?sequence=1&isAllowed=y (accessed January 25, 2020).

Fostervold, K. J., and J. Nersveen. 2008. Proportions of direct and indirect indoor lighting – the effect on health, well-being and cognitive performance of office workers. *Lighting Res Technol* 40(3):175–200. DOI: 10.1177/1477153508090917.

Fotios, S. 2017. A Revised Kruithof Graph Based on Empirical Data. *Leukos* 13(1):3–17. DOI: 10.1080/15502724.2016.1159137.

Franze, K., J. Grosche, S. N. Skatchkov et al. 2007. Muller cells are living optical fibers in the vertebrate retina. *Proc Natl Acad Sci USA* 104(20):8287–8292.

Funke, C., and C. H. Schierz. 2015. Extension of the unified glare rating formula for non-uniform led luminaires. In: *Proceedings of 28th CIE Session*, 27 June–4 July, 2015, Manchester, UK, 1471–1480.

Gall, D., and K. Bieske. 2004. Definition and measurement of circadian radiometric quantities, Light and health – non visual effects. In: *Proceedings of the CIE symposium'04, Vienna, Austria, 30 September–2 October 2004*, 129–132. https://www.db-thueringen.de/servlets/MCRFileNodeServlet/dbt_derivate_00008659/CIELuH2004_GB.pdf (accessed September 15, 2019).

Gibbs, M., S. Hampton, L. Morgan, and J. Arendt. 2002. Adaptation of the circadian rhythm of 6-sulphatoxymelatonin to a shift schedule of seven nights followed by seven days in offshore oil installation workers. *Neurosci Lett* 325(2):91–94, DOI: 10.1016/S0304-3940(02)00247-1.

Girard, J., C. Villa, and R. Brémond. 2019. Discomfort glare from several sources: A formula for outdoor lighting. *Leukos*. DOI: 10.1080/15502724.2019.1628648.

Glamox. 2019. Human centric lighting in the industry. https://glamox.com/gsx/industry1. (Accessed, September 12, 2019).

Gomes, C. C., and J. Preto. 2015. Should the light be static or dynamic? 6th International Conference on Applied Human Factors and Ergonomics (AHFE 2015) and the Affiliated Conferences, AHFE 2015. *Procedia Manuf* 3:4635–4642. DOI: 10.1016/j.promfg.2015.07.550.

Gooley, J. J, S. M. Rajaratnam, G. C. Brainard, R. E. Kronauer, C. A. Czeisler, and S. W. Lockley. 2010. Spectral responses of the human circadian system depend on the irradiance and duration of exposure to light. *Sci Transl Med* 2(31):31ra3. DOI: 10.1126/scitranslmed.3000741.

Grossberg, M. D., and S. K. Nayar. 2004. Modeling the space of camera response functions. *IEEE Trans Pattern Anal Mach Intell* 26(10):1272–1282. DOI: 10.1109/TPAMI.2004.88.

Hamedani, Z., E. Solgi, H. Skates et al. 2019. Visual discomfort and glare assessment in office environments: A review of light-induced physiological and perceptual responses. *Build Environ* 153:267–280. DOI: 10.1016/j.buildenv.2019.02.035.

Hamedani, Z., E. Solgi, T. Hine, H. Skates, G. Isoardi, and R. Fernando. 2020. Lighting for work: A study of the relationships among discomfort glare, physiological responses and visual performance. *Build Environ* 167:106478. DOI: 10.1016/j.buildenv.2019.106478.

Hamey, L. G. C. 2005. Simultaneous estimation of camera response function, target reflectance and irradiance values. In: *Proceedings of IEEE Conference Digital Image Computing: Techniques and Applications, (DICTA '05)*, 6–8 December 2005, Queensland, Australia. DOI: 10.1109/DICTA.2005.75.

Hannibal, J. 2006. Roles of PACAP-containing retinal ganglion cells in circadian timing. *Int Rev Cytol* 251:1–39. DOI: 10.1016/S0074-7696(06)51001-0.

Harmon, M. 2019. Shift work lighting. https://branchpattern.com/shift-work-lighting/ (accessed November 15, 2019).

Hart, W. M. 1987. The temporal responsiveness of vision. In: *Adler's physiology of the eye: Clinical application*, ed. R. A. Moses, and W. M. Hart. St. Louis, MI: The C. V. Mosby Company.

Hartman, P., L. Maňková, P. Hanuliak, and M. Krajčík. 2016. The influence of internal coloured surfaces on the circadian efficiency of indoor daylight. *Appl Mech Mater* 861:493–500. DOI: 10.4028/www.scientific.net/AMM.861.493.

Hattammaru, M., Y. Tahara, T. Kikuchi et al. 2019. The effect of night shift work on the expression clock genes in beard hair follicle cells. *Sleep Med* 56:164–170. DOI: 10.1016/j.sleep.2019.01.005.

Hattar, S., H. W. Liao, M. Takao, D. M. Berson, and K. W. Yau. 2002. Melanopsin-containing retinal ganglion cells: Architecture, projections, and intrinsic photosensitivity. *Science* 295(5557):1065–1070. DOI: 10.1126/science.1069609.

Higgins, K. E., and J. M. White. 1999. Transient adaptation at low light levels: Effects of age. In: *Proceedings-Vision at Low Light Levels: EPRI/LRO Fourth International Lighting Research Symposium*. Palo Alto, CA: EPRI, TR-110738, 173–185.

Hirning, M. B., G. L. Isoardi, and I. Cowling. 2014. Discomfort glare in open plan green buildings. *Energy Build* 70:427–440. DOI: 10.1016/j.enbuild.2013.11.053.

Hoekstra, J., D. P. A .Van der Groot, G. Van den Brink, and F. Bilsen. 1974. The influence of the number of cycles upon the visual contrast threshold for spatial sine wave patterns. *Vision Res* 14(6):365–368. DOI: 10.1016/0042-6989(74)90234-x.

Hoffman, G., V. Leichtfried, A. Griesmacher et al. 2008a. Effects of light with reduced short wavelength components on parameters of circadian rhythm performance in an experimental night shift model. *Open Psychol J* 1(1):34–43. DOI: 10.2174/1874360900801010034.

Hoffmann, G., V. Gufler, A. Griesmacher et al. 2008b. Effects of variable lighting intensities and colour temperatures on sulphoatoxymelatonin and subjective mood in an experimental office workplace. *Appl Ergon* 39(6):719–728. DOI: 10.1016/j.apergo.2007.11.005.

Hombach S., U. Janssen-Bienhold, G. Söhl et al. 2004. Functional expression of connexin57 in horizontal cells of the mouse retina. *Eur J Neurosci* 19(10):2633–2640. DOI: 10.1111/j.0953-816X.2004.03360.x.

Hopkinson, R. G. 1957. Evaluation of glare. *Illum Eng* 52(6):305–316.

Hopkinson, R. G. 1972. Glare from daylighting in buildings. *Appl Ergon* 3(4):206–215. DOI: 10.1016/0003-6870 (72)90102-0.

Hoshi, H., W. L. Liu, S. C. Massey, and S. L. Mills. 2009. ON inputs to the OFF layer: Bipolar cells that break the stratification rules of the retina. *J Neurosci* 29(28):8875–8883. DOI: 10.1523/JNEUROSCI.0912-09.2009.

Houser, K., M. Mossman, K. Smet, and L. Whitehead. 2016. Tutorial: Color Rendering and Its Applications in Lighting. *Leukos* 12(1–2):7–26. DOI: 10.1080/15502724.2014.989802.

https://standard.wellcertified.com/ (accessed October 2, 2019).

ICNIRP [International Commission on Non-Ionizing Radiation Protection]. 2013. ICNIRP guidelines on limits of exposure to incoherent visible and infrared radiation. *Health Phys* 105(1):74–96.

IEC 61000-4-15. 2010. Electromagnetic Compatibility (EMC)—Parts 4–15: Testing and measurement techniques—flicker meter—functional and design specifications.

IEEE 1789: 2015. IEEE recommended practices for modulating current in high-brightness LEDs for mitigating health risks to viewers.

Irikura, T., Y. Toyofuku, and Y. Aoki. 1999. Recovery time of visual acuity after exposure to a glare source. *Lighting Res Technol* 31(2):57–61.

ISO 2721. 1982. Photography – cameras – automatic controls of exposure.

ISO 8995-1. 2002. Lighting of work places - Part 1: Indoor.

Itani, O., Y. Kaneita, M. Tokiya et al. 2017. Short sleep duration, shift work, and actual days taken off work are predictive life-style risk factors for new-onset metabolic syndrome: A seven-year cohort study of 40,000 male workers. *Sleep Med* 39:87–94. DOI: 10.1016/j.sleep.2017.07.027.

Izdebski, Ł., and D. Sawicki. 2016. Easing functions in the new form based on Bézier curves. In: *Computer vision and graphics. ICCVG 2016. Lecture notes in computer science*, ed. L. Chmielewski, A. Datta, R. Kozera, and K. Wojciechowski, vol. 9972. Cham: Springer. DOI: 10.1007/978-3-319-46418-3_4.

Jaadane, I., P. Boulenguez, S. Chahory et al. 2015. Retinal damage induced by commercial light emitting diodes (LEDs). *Free Radical Bio Med* 84:373–384. DOI: 10.1016/j.freeradbiomed.2015.03.034.

Jacobson, R. E., S. F. Ray, G. G Attridge, and N. R. Axford. 2000. *Manual of photography, photographic and digital imaging*. 9th ed. Oxford, UK: Focal Press,.

Jarboe, C., J. Snyder, and M. G. Figueiro. 2019. The effectiveness of light emitting diode lighting for providing circadian stimulus in office spaces while minimizing energy use. *Lighting Research Technol* 1–22. DOI: 10.1177/1477153519834604.

Jerath, R, S. M. Cearley, V. A.Barnes, and E. Nixon-Shapiro. 2016. How lateral inhibition and fast retinogeniculo-cortical oscillations create vision: A new hypothesis. *Med Hypotheses* 96:20–29. DOI: 10.1016/j.mehy.2016.09.015.

Jordan, G., and J. D. Mollon. 1993. A study of women heterozygous for colour deficiencies. *Vision Res* 33(11):1495–1508.

Jung, B., and M. Inanici. 2018. Measuring circadian lighting through high dynamic range photography. *Lighting Research Technol* 51(5):742–763. DOI: 10.1177/1477153518792597.

Juslen, H. 2007. Lighting, productivity and preferred illuminances – field studies in the industrial environment. PhD Thesis. Helsinki: Helsinki University of Technology.

Juslen, H. M. Wouters , and A. Tenner. 2007. The influence of controllable task-lighting on productivity: A field study in a factory. *Appl Ergon* 38(1):39–44. DOI: 10.1016/j.apergo.2006.01.005.

Kakitsuba, N. 2020. Comfortable indoor lighting conditions for LED lights evaluated from psychological and physiological responses. *Appl Ergon* 82:102941. DOI: 10.1016/j.apergo.2019.102941.

Kanagawa, M, Y. Omori, S. Sato et al. 2010. Post-translational maturation of dystroglycan is necessary for pikachurin binding and ribbon synaptic localization. *J Biol Chem* 285(41):31208–31216. DOI: 10.1074/jbc.M110.116343.

Kang, H. R. 2006. *Computational color technology*. Bellingham: SPIE Press.

Kayumov, L., R. F. Casper, R. J. Hawa et al. 2005. Blocking low wavelength light prevents noc-turnal melatonin suppression with no adverse effect on performance during simulated shift work, *J Clin Endocrinol Metab* 90(5):2755–2761. DOI: 10.1210/jc.2004-2062.

Kearney, A. T. 2015. Quantified benefits of Human Centric Lighting, Lighting Europe. https://www.lightingeurope.org/images/publications/presentations/150420_From_Bar riers_to_Measures_-_Final_Results_-_Complete_vF_004.pdf (accessed October 17, 2019).

Kelly, D. H. 1961. Visual responses to time-dependent stimuli. I. Amplitude sensitivity mea-surements. *J Opt Soc Am* 51(4):422–429. DOI: 10.1364/JOSA.51.000422.

Khanh, T. Q, P. Bodrogi, Q. T. Vinh, and H. Winkler. 2015. *LED lighting: Technology and perception*, Weinheim: Wiley-VCH Verlag GmbH & Co KGaA.

Kim, J. B., and J. H. Kim. 2017. Regional gray matter changes in shift workers: A voxel-based morphometry study. *Sleep Med* 30:185–188. DOI: 10.1016/j.sleep.2016.10.013.

Klawe, J. J., A. Laudencka, I. Miśkowiec, and M. Tafil-Klawe. 2005. Occurrence of obstruc-tive sleep apnea in a group of shift worked police officers. *J Physiol Pharmacol* 56(4): 115–117.

Knoop, M., K. Broszio, A. Diakite et al. 2019. Methods to describe and measure lighting con-ditions in experiments on non-image-forming aspects, *Leukos* 15(2–3):163–179. DOI: 10.1080/15502724.2018.1518716.

Knutsson, A., and A. Kompa. 2014. Shift work and diabetes — a systematic review. *Chronobiol Int* 31(10):1146–1151. DOI: 10.3109/07420528.2014.957308.

Kolb, H., E. Fernandez, and R. Nelson. 1995. *The Organization of the retina and visual sys-tem*. Salt Lake City, UT: Webvision.

Kombeiz, O., and A. Steidle. 2018. Facilitation of creative performance by using blue and red accent lighting in work and learning areas. *Ergonomics* 61(3):456–463. DOI: 10.1080/00140139.2017.1349940.

Konis, K. 2017. A novel circadian daylight metric for building design and evaluation. *Build Environ* 113:22–38. DOI: 10.1016/j.buildenv.2016.11.025.

Koo Y., J. Choi, and K. Jung. 2013. Sleep disturbances and their relationship with excessive exposure to light at night: The Korean genome and epidemiology study. *Sleep Med* 14(1):e29–e30.

Krigel, A., M. Berdugo, E. Picard et al. 2016. Light-induced retinal damage using different light sources, protocols and rat strains reveals led phototoxicity. *Neuroscience* 339:296–307. DOI: 10.1016/j.neuroscience.2016.10.015.

Kruithof, A. A. 1941. Tubular Luminescence Lamps for General Illumination. *Philips Technical Review* 6(3):65–96.

Ksendzovsky, A., I. J. Pomeraniec, K. A. Zaghloul, J. J. Provencio, and I. Provencio. 2017. Clinical implications of the melanopsin-based non-image-forming visual system. *Neurology* 88(13):1282–1290. DOI: 10.1212/WNL.0000000000003761.

Lal, S. K. I., and A. Craig. 2001. A critical review of the psychophysiology of driver fatigue. *Biol Psychol* 55(3):173–194. DOI: 10.1016/S0301-0511(00)00085-5.

Lasance, C. J. M., and A. Poppe. 2014. *Thermal management for LED applications*. New York: Springer.

Łaszewska, K., A. Goroncy, P. Weber et al. 2017. Daytime acute non-visual alerting response in brain activity occurs as a result of short- and long wavelengths of light. *J Psychophysiol* 32(4):202–226. DOI: 10.1027/0269-8803/a000199.

Łaszewska K., A. Goroncy, P. Weber, T. Pracki, and M. Tafil-Klawe.2018. Influence of the spectral quality of light on daytime alertness levels in humans. *Adv Cognit Psychol* 14(4):192–208. DOI: 10.5709/acp-0250-0.

Lavoie, S., J. Paquet, B. Selamoui, M. Rufiange, and M. Dumont. 2003. Vigilance levels during and after bright light exposure in the first half of the night. *Chronobiol Int* 20(6):1019–1038. DOI: 10.1081/CBI-120025534.

Lee, J., J. Kim, and M. Shin. 2017. Correlation analysis between electrocardiography (ECG) and photoplethysmogram (PPG) data for driver's drowsiness detection using noise replacement method. *Procedia Comp Sci* 116:421–426. DOI: 10.1016/j.procs.2017.10.083.

Lee, J-H., J. W. Moon, and S. Kim. 2014. Analysis of occupants' visual perception to refine indoor lighting environment for office tasks. *Energies* 7:4116–4139. DOI: 10.3390/en7074116.

Lehnert, P. 2001. Disability and discomfort glare under dynamic conditions - the effect of glare stimuli on the human vision. In: *Proceedings of PAL 2001 - Progress in Automobile Lighting, Held Laboratory of Lighting Technology*, Darmstadt, Germany, September 2001, vol. 9, 582–592.

licht. wissen 19. 2014. Impact of light on human beings. https://www.licht.de/fileadmin/Publ ications/licht-wissen/1409_LW19_E_Impact-of-Light-on-Human-Beings_web.pdf (accessed January 20, 2020).

licht. wissen 21. 2018. Guide to human centric lighting (HCL). https://en.licht.de/fileadmin/ Publications/licht-wissen/1809_lw21_E_Guide_HCL_web.pdf (accessed January 20, 2020).

LHA [Light and Health Alliance]. 2019. Light and shift workers. Lighting patterns for healthy building. http://lightingpatternsforhealthybuildings.org/content/19 (accessed September 12, 2019).

Lin, Y., S. Fotios, M. Wei, Y. Liu, W. Guo, and Y. Sun. 2015. Eye movement and pupil size constriction under discomfort glare. *Investig Ophthalmol Vis Sci* 56:1649–1656. DOI: 10.1167/iovs.14-15963.

Lipetz, L. E. 1971. The relation of physiological and psychological aspects of sensory intensity. In: *Principles of receptor physiology*, ed. W. R. Loewestein, vol. 1. Heidelberg: Springer, 191–225.

Lodetti, S., I. Azcarate, J. J. Gutiérrez et al. 2019. Flicker of modern lighting technologies due to rapid voltage changes. *Energies* 12:865–880. DOI: 10.3390/en12050865.

Lowry, G. A. 2018. Comparison of metrics proposed for circadian lighting and the criterion adopted in the WELL Building Standard. In: *Proceedings of CIBSE Technical Symposium*, London UK, 12–13 April 2018, 1–10. https://openresearch.lsbu.ac.uk/dow nload/40bab6a72fef47a72ee17721b2f1f407fbc6ac3eafbcae6eb2588f297b02dc39/3617 53/032%20final.pdf (accessed November 10, 2019).

LRC [Lighting Research Center]. 2018a. Lighting for health and energy saving: Human factors. https://www.lrc.rpi.edu/resources/newsroom/pdf/2018/Lighting_for_Health_Ene rgy_Savings_8511.pdf (accessed October 5, 2019).

LRC [Lighting Research Center]. 2018b. Technical report: Increasing circadian light exposure in office spaces. https://www.lrc.rpi.edu/programs/lightHealth/pdf/GSA/DOS_2018.pdf (accessed October 5, 2019).

Lucas Group. 2019. Measuring melanopic illuminance, The Lucas Group Website in the World Wide Web. The University of Manchester. http://lucasgroup.lab.manchester.ac.uk /measuringmelanopicilluminance/ (accessed November 10, 2019).

Lucas, R., S. Peirson, D. Berson et al. 2014. Measuring and using light in the melanopsin age. *Trends Neurosci* 37(1):1–9. DOI: 10.1016/j.tins.2013.10.004

Luckiesh, M., and S. K. Guth. 1949. Brightness in visual field at borderline between comfort and discomfort (BCD). *Illum Eng* 44(11):650–670.

Mannos, J. L., and D.J. Sakrison. 1974. The effects of a visual fidelity criterion on the encoding of images. *IEEE Trans Inf Theory* 20(4):525–536. DOI: 10.1109/TIT.1974.1055250.

Manov, B. 2007. An experimental study on the appraisal of the visual environment at offices in relation to colour temperature and illuminance. *Build Environ* 42(2):979–983. DOI: 10.1016/j.buildenv.2005.10.022.

Mantiuk, R. K., K. Myszkowski, and H-P. Seidel. 2015. High dynamic range imaging. In: *Wiley encyclopedia of electrical and electronics engineering*. John Wiley & Sons, Inc. DOI: 10.1002/047134608X.W8265.

Marek, V., E. Reboussin, J. Degardin-Chicaud et al. 2019. Implication of melanopsin and trigeminal neural pathways in blue light photosensitivity in vivo. *Front Neurosci* 13:497. DOI: 10.3389/fnins.2019.00497.

Matthews, G. G. 1998. *Neurobiology: Molecules, cells, and systems.* Maiden: Blackwell Science.

Matynia, A., E. Nguyen, X. Sun et al. 2016. Peripheral sensory neurons expressing melanopsin respond to light. *Front Neural Circuits* 10:60. DOI: 10.3389/fncir.2016.00060.

Mayeli, M. 2019. Neurophysiology of visual perception. In: *Biophysics and neurophysiology of the sixth sense*, ed. N. Rezaei, and A. Saghazadeh. Springer, 13–26. DOI: 10.1007/978-3-030-10620-1_2.

Merbs, S. L., and J. Nathans. 1992. Absorption spectra of human cone pigments. *Nature.* 356(6368):433–435.

Michelson, A. 1927. *Studies in optics.* Chicago, IL: The University of Chicago Press.

Mokrzycki, W. S., and M. Tatol. 2011. Color difference ΔE: A survey. *Machine Graphics and Vision* 20(4), 383–411.

Molday, R. S., and O. L. Moritz. 2015. Photoreceptors at a glance. *J Cell Sci* 128(22):4039–4045. DOI: 10.1242/jcs.175687.

Mollon, J. D., J. Pokorny, and K. Knoblauch. 2001. *Normal and defective colour vision.* Oxford, UK: Oxford University Press.

Moore-Ede, M. C., A. Heitmann, and R. G. Guttkuhn. 2020. Circadian potency spectrum with extended exposure to polychromatic white LED light under workplace conditions. *J Biol Rhythms*, In Press, DOI: 10.1177/0748730420923164.

Moosmann, C., J. Wienold, A. Wagner, and V. Wittwer. 2009. Age effects on glare perception under daylight conditions. In: *Proceedings of Conference LUX EUROPA 2009*, Istanbul, 439–442.

Mott, M. S., D. H. Robinson, A. Walden, J. Burnette, and A. S. Rutherford. 2012. Illuminating the effects of dynamic lighting on student learning. *SAGE Open* 2(2):1–9, DOI: 10.1177/2158244012445585.

Moulton, E. A., L. Becerra, and D. Borsook. 2009. An fMRI case report of photophobia: Activation of the trigeminal nociceptive pathway. *Pain* 145(3):358–363. DOI: 10.1016/j.pain.2009.07.018.

Nagare, R. 2017. The effect of the spectral content of a glare source and its background on discomfort glare. http://cormusa.org/wp-content/uploads/2018/04/03-01_RNagare.pdf (accessed December 15, 2019).

Nagare, R., B. Plitnick, and M. G. Figueiro. 2019a. Effect of exposure duration and light spectra on nighttime melatonin suppression in adolescents and adults. *Lighting Res Technol* 1–14. DOI: 10.1177/1477153518763003.

Nagare, R., M. S. Rea, B. Plitnick, and M. G. Figueiro. 2019b. Nocturnal Melatonin suppression by adolescents and adults for different levels, spectra, and durations of light exposure. *J Biol Rhythms* 34(2):178–194. DOI: 10.1177/0748730419828056.

NEMA 77-2017. 2017. Temporal light artifacts: Test methods and guidance for acceptance criteria.

Nicholls, J. G., A. R. Martin, B. G. Wallace, and P. A. Fuchs. 2001. *From neuron to brain: A cellular and molecular approach to the function of the nervous system.* 4th ed. Sunderland: Sinauer Associates.

NOAO [National Optical Astronomy Observatory]. 2019. Recommended light levels. https://www.noao.edu/education/QLTkit/ACTIVITY_Documents/Safety/LightLevels_outdoor+indoor.pdf) (accessed September 10, 2019).

Nonne, J., D. Renoux, and L. Rossi. 2013. Metrology for solid-state lighting quality. In: *Proceedings of 16th International Congress of Metrology*, October 1–10, Paris, France, 14004. DOI: 10.1051/metrology/201314004.

Ohayon, M. M., and C. Milesi. 2016. Artificial outdoor nighttime lights associate with altered sleep behavior in the American general population. *Sleep* 39(6): 1311–1320. DOI: 10.5665/sleep.5860.

Okamoto, Y., M. S Rea, and M. G. Figueiro. 2014. Temporal dynamics of EEG activity during short- and long-wavelength light exposures in the early morning. *BMC Res Notes* 7:113. DOI: 10.1186/1756-0500-7-113.

Omori, Y, F. Araki, T. Chaya et al. 2012. Presynaptic dystroglycan-pikachurin complex regulates the proper synaptic connection between retinal photoreceptor and bipolar cells. *J Neurosci* 32(18):6126–6137. DOI: 10.1523/JNEUROSCI.0322-12.2012.

OpenEXR. 2019. Academy software foundation. http://www.openexr.com/ (accessed December 12, 2019).

Ortega, J. T., T. Parmar, and B. Jastrzebska. 2019. Flavonoids enhance rod opsin stability, folding, and self-association by directly binding to ligand-free opsin and modulating its conformation. *J Biol Chem* 294(20):8101–8122. DOI: 10.1074/jbc.RA119.007808.

Palmer, C. A., and C. A. Alfano. 2017. Sleep and emotion regulation: An organizing, integrative review. *Sleep Med Rev* 31:6–16. DOI: 10.1016/j.smrv.2015.12.006.

Pang, J. J., F. Gao, J. Lem, D. E. Bramblett, D. L. Paul, and S. M. Wu. 2010. Direct rod input to cone BCs and direct cone input to rod BCs challenge the traditional view of mammalian BC circuitry. *Proc Natl Acad Sci USA* 107(1):395–400.

Pechacek, C. S., M. Anderson, and S. W. Lockley. 2008. Preliminary method for prospective analysis of the circadian efficacy of (day) light applications to healthcare architecture. *Leukos* 5(1):1–26. DOI: 10.1582/LEUKOS.2008.05.01001.

Pedler, D., and M. G. Figueiro. 2017. Lighting focused on occupant health and well-being, *Building Operating Management*. https://www.facilitiesnet.com/lighting/article.aspx?id=17353&source=part_o (accessed October 12, 2019).

Penner, R. 2002. Motion, tweening, and easing. In: *Programming macromedia Flash MX*. McGraw-Hill/OsborneMedia, 191–240. http://robertpenner.com/easing/penner_chapter 7_tweening.pdf (accessed October 6, 2019).

Perz, M. 2019. Modelling visibility of temporal light artefacts. PhD thesis. Eindhoven: Eindhoven University of Technology.

Pharr, M., J. Wenzel, and G. Humphreys. 2016. *Physically based rendering: From theory to implementation*. 3rd ed. Cambridge, MA: Morgan Kaufmann is an imprint of Elsevier.

Phipps-Nelson, J., J. R. Redman, L. J. Schlangen, and S. M. Rajaratnam. 2009. Blue light exposure reduces objective measures of sleepiness during prolonged night time performance testing. *Chronobiol Int* 26(5):891–912. DOI: 10.1080/0742052090304 4364.

Pi Lighting. 2017. Paradigm changes in lighting R&D 2017 changed. https://www.swisspho tonics.net/libraries.files/Bataillou.pdf.

Pickard, G. E., S. B. Baver, M. D. Ogilvie, and P. J. Sollars. 2009. Light-induced fos expression in intrinsically photosensitive retinal ganglion cells in melanopsin knockout (opn4) mice. *PloS One* 4(3):e4984. DOI: 10.1371/journal.pone.0004984.

Pickard, G. E., and P. Sollars. 2012. Intrinsically photosensitive retinal ganglion cells. *Rev Physiol Biochem Pharmacol* 162:59–90. DOI: 10.1007/112_2011_4.

Pieh, C., R. Jank, C. Waiss et al. 2018. Night-shift work increases cold pain perception. *Sleep Med* 45:74–79. DOI: 10.1016/j.sleep.2017.12.014.

Preto, S., and C. C. Gomes. 2019. Lighting in the workplace: Recommended illuminance (lux) at workplace environs. In: *Advances in Design for Inclusion*, ed. G. Di Bucchianico, AHFE 2018. AISC 776, Springer International Publishing AG, 180–191. DOI: 10.1007/978-3-319-94622-1_18.

Price, L. L. A., and A. Lyachev. 2017a. Research note: Modification of a personal dosimetry device for logging melanopic irradiance. *Lighting Res Technol* 49: 922–927. DOI: 10.1177/1477153517695862.

Price, L. L. A., A. Lyachev, and M. Khazova. 2017b. Optical performance characterization of light-logging actigraphy dosimeters, *J Opt Soc Am A Opt Image Sci Vis* 34(4):545–557. DOI: 10.1364/JOSAA.34.000545.

Provencio, I., I. R. Rodriguez, G. Jiang, W. P. Hayes, E. F. Moreira, and M. D. Rollag. 2000. A novel human opsin in the inner retina. *J Neurosci* 20(2):600–605.

Provencio, I., M. D. Rollag, and A. M. Castrucci. 2002. Photoreceptive net in the mammalian retina: This mesh of cells may explain how some blind mice can still tell day from night. *Nature* 415(6871):493.

Pulpitlova J., and J. Detkova. 1993. Impact of the cultural and social background on the visual perception in living and working perception. In: *Proceedings of the International Symposium Design of Amenity*, October 5–9, 1993, Fukuoka, Japan, 216–227.

Purves, D., and R. B. Lotto. 2003. *Why we see what we do: an empirical theory of vision*. Sunderland, MA: Sinuaer Associates Inc.

Radiance. 2019. A validated lighting simulation tool. RADSITE radiance-online.org. https://www.radiance-online.org/ (accessed December 12, 2019).

RADSITE [radiance-online.org]. 2019. https://www.radiance-online.org/learning/documenta tion/manual-pages/pdfs/evalglare.pdf/view (accessed November 25, 2019).

Rahman, S. A., S. Marcu, C. M. Shapiro, T. J Brown, and R. F. Casper. 2011. Spectral modulation attenuates molecular, endocrine, and neurobehavioral disruption induced by nocturnal light exposure. *Am J Physiol Endocrinol Metab* 300(3):518–527. DOI: 10.1152/ajpendo.00597.2010.

Rahman, S. A., C. M. Shapiro, F. Wang et al. 2013. Effects of filtering short wavelengths during nocturnal shift work on sleep and performance. *Chronobiol Int* 30(8), 951–962. DOI: 10.3109/07420528.2013.789894.

Rahman, S. A., E. E Flynn-Evans, D. Aeschbach, G. C. Brainard, C. A. Czeisler, and S. W. Lockley. 2014. Diurnal spectral sensitivity of the acute alerting effects of light. *Sleep* 37(2):271–281. DOI: 10.5665/sleep.3396.

Ray, S. F. 2002. *Applied photographic optics*. Oxford, UK: Focal Press.

Rea, M. S. 1981. Visual performance with realistic methods of changing contrast. *J Illum Eng Soc* 10(3):164–177. DOI: 10.1080/00994480.1980.10748607.

Rea, M. S. 1986. Toward a model of visual performance: Foundations and data. *J Illum Eng Soc* 15(2):41–57. DOI: 10.1080/00994480.1986.10748655

Rea, M. S., and M. J. Ouellette. 1988. Visual performance using reaction times. *Lighting Res Technol* 20(4):139–153. DOI: 10.1177/096032718802000401.

Rea, M. S., and M. J. Ouellette. 1991. Relative visual performance: A basis for application. *Lighting Res Technol* 23(3):135–144. DOI: 10.1177/096032719102300301.

Rea, M. S. 2000. *The IESNA lighting handbook: Reference and application*. New York: Illuminating Engineering Society of North America.

Rea, M. S., J. D. Bullough, M. G. Figueiro, and A. Bierman. 2004. Spectral opponency in human circadian phototransduction: Implications for lighting practice. In: *Proceedings of the CIE symposium 2004*, Vienna, Austria, 30 September–2 October 2004, 111–115.

Rea, M. S., M. G. Figueiro, J. D. Bullough, and A. Bierman. 2005. A model of phototransduction by the human circadian system. *Brain Res Rev* 50: 213–228. DOI: 10.1016/j.brainresrev.2005.07.002.

Rea, M. S, and J. P. Freyssinier-Nova. 2008. Color rendering: A tale of two metrics. *Color Res Appl* 33(3):192–202. DOI: 10.1002/col.20399.

Rea, M. S., M. G. Figueiro, A. Bierman, and J. D. Bullough. 2010. Circadian light. *J Circadian Rhythms* 8(1):2. DOI: 10.1186/1740-3391-8-2.

Rea, M. S., M. G. Figueiro, A. Bierman, and R. Hamner. 2012. Modelling the spectral sensitivity of the human circadian system. *Lighting Res Technol* 44(4):386–396. DOI: 10.1177/1477153511430474.

Rea, M. S. 2015. The lumen seen in a new light: Making distinctions between light, lighting and neuroscience. *Lighting Res Technol* 47:259–280. DOI: 10.1177/1477153514527599.

Rea, M. S., and M. G. Figueiro. 2016. Light as a circadian stimulus for architectural lighting. *Lighting ReS Technol* 50(4):497510. DOI: 10.1177/1477153516682368.

Rea, M. S. 2018. Lighting simply made better: Providing a full range of benefits without much fuss. *Build Environ* 144:57–65. DOI: 10.1016/j.buildenv.2018.07.047.

Reinberg, A., and I. Ashkenazi. 2008. Internal desynchronization of circadian rhythms and tolerance to shift work. *Chronobiol Int* 25(4):625–643. DOI: 10.1080/07420520802256101.

Reinhard, E., G. Ward, S. Pattanaik, and P. Debevec. 2006. *High dynamic range imaging*. San Francisco, CA: Morgan Kaufmann.

Relux. 2019. Lighting design software. https://relux.com/en/ (accessed December 19, 2019).

Roecklein, K. A., K. J. Rohan, W. C. Duncan et al. 2009. A missense variant (P10L) of the melanopsin (OPN4) gene in seasonal affective disorder. *J Affect Disord* 114(1–3):279–285. DOI: 10.1016/j.jad.2008.08.005.5.

Roenneberg, T., and M. Merrow. 2016. The circadian clock and human health. *Current Biology* 26(10):432–443. DOI: 10.1016/j.cub.2016.04.011.

Rosenhahn, E., and M. Lampen. 2004. New investigation of the subjective glare effect of projection type headlamps. *SAE Technical Paper*. 2004-01-1281. DOI: 10.4271/2004-01-1281.

Rossi, M. 2019. *Circadian lighting design in the LED area*. Cham, Switzerland: Springer Nature.

Royer, M. P. 2018. Comparing measures of average color fidelity. *Leukos* 14(2):69–85. DOI: 10.1080/15502724.2017.1389283.

Safdar, M., M. R. Luo, M. F. Mughal et al. 2018. A neural response-based model to predict discomfort glare from luminance image. *Lighting Res Technol* 50(3):416–428. DOI: 10.1177/1477153516675910.

Sahin, L., and M. G. Figueiro. 2013. Alerting effects of short-wavelengths (blue) and long-wavelengths (red) lights in the afternoon. *Physiol Behav* 116:1–7. DOI: 10.1016/j.physbeh.2013.03.014.

Sahin, L., B. M. Wood, M. Plitnick, and M. G. Figueiro. 2014. Daytime light exposure: Effects on biomarkers, measures of alertness, and performance. *Behav Brain Res* 274:176–185. DOI: 10.1016/j.bbr.2014.08.017.

Saksvik, I. B., B. Bjorvatn, H. Hetland, G. M. Sandal, and S. Pallesen. 2011. Individual differences in tolerance to shift work-a systematic review. *Sleep Med Rev* 15(4):221–235. DOI: 10.1016/j.smrv.2010.07.002.

Saper, C. B., P. M. Fuller, N. P. Pedersen, J. Lu, and T. E. Scammell. 2010. Sleep state switching. *Neuron* 68(6):1023–1042. DOI: 10.1016/j.neuron.2010.11.032.

Sawicki, D., and A. Wolska. 2013. Algorithm of HDR image preparation for discomfort glare assessment. *Przegląd Elektrotechniczny* 89(2a):87–90.

Sawicki, D., and A. Wolska. 2015. Discomfort glare prediction by different methods. *Lighting Res Technol* 47(6):658–671. DOI: 10.1177/1477153515589773.

Sawicki, D., and A. Wolska. 2016a. The unified semantic glare scale for GR and UGR indexes. In: *Proceedings of the 2016 IEEE Lighting Conference of the Visegrad Countries Karpacz*, 13–16 September, 2016, Poland. DOI: 10.1109/LUMENV.2016.7745536.

Sawicki, D., and A. Wolska. 2016b. Problems related to the angular resolution of the ILMD for GR index determination. *Przegląd Elektrotechniczny* 92(9):173–177. DOI: 10.15199/48.2016.09.44.

Sawicki, D., A. Wolska, P. Rosłon, and S. Ordysiński. 2016c. New EEG measure of the alertness analyzed by Emotiv EPOC in a real working environment. In: *Proceedings of the 4th International Congress on Neurotechnology, Electronics and Informatics*, 7–8 November 2016, Porto, Portugal, 35–42. DOI: 10.5220/0006041200350042.

Sawicki, D., A. Wolska, M. Wisełka, and S. Ordysiński. 2019. Easing function as a tool of color correction for display stitching in virtual reality In: *Proceedings of the 20th International Conference on Image Analysis and Processing (ICIAP 2019)*. DOI: 10.1007/978-3-319-68548-9_33.

Scheir, G. H., P. Hanselaer, P. Bracke, G. Deconinck, and W. R. Ryckaert. 2015. Calculation of the Unified Glare Rating based on luminance maps for uniform and non-uniform light sources. *Build Environ* 84:60–67. DOI: 10.1016/j.buildenv.2014.10.027.

Scheir, G. H., M. Donners, L. M. Geerdinck et al. 2018. A psychophysical model for visual discomfort based on receptive fields. *Lighting Res Technol* 50:205–217. DOI: 10.1177/1477153516660606.

Scheir, G. H., P. Hanselaer, and W. R. Ryckaert. 2019. Pupillary light reflex, receptive field mechanism and correction for retinal position for the assessment of visual discomfort. *Lighting Res Technol* 51(2):291–303. DOI: 10.1177/1477153517737346.

Schen, Y., C. Xie, Y. Gu, X. Li, and J. Tong. 2016. Illumination from light-emitting diodes (LEDs) disrupts pathological cytokines expression and activates relevant signal pathways in primary human retinal pigment epithelial cells. *Exp Eye Res* 145:456–467. DOI: 10.1016/j.exer.2015.09.016.

Scheuermaier, K., M. Munch, J. M. Ronda, and J. F. Duffy. 2018. Improved cognitive morning performance in healthy older adults following blue-enriched light exposure on the previous evening. *Behav Brain Res* 348:267–275, DOI: 10.1016/j.bbr.2018.04.021.

Schieber, F. 1992. Aging and the Senses. In: *Handbook of mental health and aging*. ed. J. E. Birren, G. D. Cohen, R. B. Sloane et al. San Diego, CA: Academic Press.

Schierz, C, and C. Vandahl. [2019]. Biological effects of light – Literature overview. https i//www.google.com/search?q=12.%09Schierz i C%2C i Vandahl i C. i Biological i effects +of+light+%E2%80%93+Literature+overview&ie=utf-8&oe=utf-8&client=firefox-b# (accessed October 20, 2019).

Schiller, P. H. 1992. The ON and OFF channels of the visual system. *Trends Neurosci* 15(3):86–92. DOI: 10.1016/0166-2236(92)90017-3.

Schiller, P. H. 2010. Parallel information processing channels created in the retina. *Proc Natl Acad Sci U S A* 107(40):17087–17094. DOI: 10.1073/pnas.1011782107.

Schlangen, L. 2016. Human Centric Lighting needs new quantities for light intensity. *Light + Building, Frankfurt, March 15th 2016*. http://lightingforpeople.eu/2016/wp-content/u ploads/2015/08/HCL-needs-new-quantities-for-light-intensity.pdf.

Schlesselman, B., M. Gordin, L. Boxler et al. 2015. Brightness judgment in a simulated sports field correlate with S/P value of light sources. In: *Proceedings of the IES National Conference,* Indianapolis IN, 8–10 November, 2015, New York, 1–18.

Schmidt, T. M., and P. Kofuji. 2010. Differential cone pathway influence on intrinsically photosensitive retinal ganglion cell subtypes. *J Neurosci* 30(48):16262–16271. DOI: 10.1523/JNEUROSCI.3656-10.2010.

Schmidt, T., S. K. Chen, and S. Hattar. 2011a. Intrinsically photosensitive retinal ganglion cells: Many subtypes, diverse functions. *Trends Neurosci* 34:572–580. DOI: 10.1016/j. tins.2011.07.001.

Schmidt, T. M., M. T. Do, D. Dacey et al. 2011b. Melanopsin-positive intrinsically photosensitive retinal ganglion cells: From form to function. *J Neurosci* 31(45):16094–16101. DOI: 10.1523/JNEUROSCI.4132-11.2011.

Schmidt, T. M., and P. Kofuji. 2011c. Structure and function of bistratified intrinsically photosensitive retinal ganglion cells in the mouse. *J Comp Neurol* 519(8):1492–1504. DOI: 10.1002/cne.22579.

Schwitzer, T., R. Schwan, E. Albuisson et al. 2017a. Association between regular cannabis use and ganglion cell dysfunction. *JAMA Ophtalmol* 135(1) 54–60. DOI: 10.1037/ abn0000347.

Schwitzer, T., R. Schwan, F. Bubl et al. 2017b. Looking into the brain through the retinal ganglion cells in psychiatric disorder: A review of evidences. *Prog Neuro-Psychopharmacol Biol Psychiatry* 76:155–162. DOI: 10.1016/j.pnpbp.2017.03.008.

Schwitzer, T., R. Schwan, K. Angioi-Duprez et al. 2018. Delayed bipolar and ganglion cells neuroretinal processing in regular cannabis users: The retina as a relevant site to investigate brain synaptic transmission dysfunction. *J Psychiatr Res* 103:75–82. DOI: 10.1016/j.jpsychires.2018.04.021.

Semo, M., C. Gias, A. Ahmado, and A. Vugler. 2014. A role for the ciliary marginal zone in the melanopsin-dependent intrinsic pupillary light reflex. *Exp Eye Res* 119:8–18. DOI: 10.1016/j.exer.2013.11.013.

Setchell, J. S. 2012. Colour description and communication. In: *Colour design: Theories and applications.* ed J. Best, 219–253. DOI: 10.1533/9780857095534.2.219.

Shekhar, K., S. W. Lapan, I. E. Whitney et al. 2016. Comprehensive classification of retinal bipolar neurons by single-cell transcriptomics. *Cell* 166(5):1308–2330. DOI: 10.1016/j.cell.2016.07.054.

Shelley, J. 2006. Horizontal cel receptive fields are reduced in connexin 5–7 deficient mice. *Eur J Neurosci* 23(12):3176–3186. DOI: 10.1111/j.1460-9568.2006.04848.x.

Sithravel RK, Ibrahim R, Lye MS et al. 2018. Morning boost on individuals' psychophysiological wellbeing indicators with supportive, dynamic lighting in windowless open-plan workplace in Malaysia. *PLoS One* 13(11):e0207488.

Smith, T., and J. Guild. 1931–1932. The C.I.E. colorimetric standards and their use. *Trans Opt Soc* 33(3):73–134. DOI: 10.1088/1475-4878/33/3/301.

Smolders, K. C. H. J., and Y. A. W. de Kort. 2009. Light up my day: A between-group test of dynamic lighting effects on office workers' wellbeing in the field. In: *Proceedings of 11th European Lighting Congress (Lux Europa 2009)*, September 9–11, 2009, Istanbul, Turkey, 161–168.

Smolders, K. C. J. H., Y. A. W. de Kort, and P. J. M. Cluitmans. 2012. A higher illuminance induces alertness even during office hours: Findings on subjective measures, task performance and heart rate measures. *Physiol Behav* 107(1):7–16. DOI: 10.1016/j.physbeh.2012.04.028.

Smolders, K. C. H. J., Y. A. W. de Kort, and S. M. van den Berg. 2013. Daytime light exposure and feelings of vitality: Results of a field study during regular weekdays. *J Environ Psychol* 36:270–279. DOI: 10.1016/j.envp.2013.09.004.

Snowden, R., P. Thompson, and T. Troscianko. 2012. *Basic vision: An introduction to visual perception.* Oxford, UK: Oxford University Press.

Soler, R. 2019. Circadian rhythms. https://resources.wellcertified.com/articles/circadian-rhythms/ (accessed November 15, 2019).

Sollars, P. J., and G. E. Pickard. 2015. The neurobiology of circadian rhythms. *Psychiatr Clin North Am* 38(4):645–665. DOI: 10.1016/j.psc.2015.07.003.

Sonoda, T., and T. M. Schmidt. 2016. Re-evaluating the role of intrinsically photosensitive retinal ganglion cells: New roles in image-forming functions. *Integr Comp Biol* 56(5):834–841. DOI: 10.1093/icb/icw066.

Spitschan, M., and G. K. Aguirre. 2017a. Vision: Melanopsin as a raumgeber. *Curr Biol* 27(13):644–646. DOI: 10.1016/j.cub.2017.05.052.

Spitschan, M., A. S. Bock, J. Ryan et al. 2017b. The human visual cortex response to melanopsin-directed stimulation is accompanied by a distinct perceptual experience. *Proc Nat Acad Sci U S A* 114(46):12291–12296. DOI: 10.1073/pnas.1711522114.

Spitschan, M., R. J. Lucas, and T. M. Brown. 2017c. Chromatic clocks: Color opponency in non-image-forming visual function. *Neurosci Biobehav Rev* 78:24–33. DOI: 10.1016/j.neubiorev.2017.04.016.

Spitschan, M. 2019a. Melanopsin contributions to non-visual and visual function. *Curr Opin Beha Sci* 30:67–72. DOI: 10.1016/j.cobeha.2019.06.004.

Spitschan, M. 2019b. Photoreceptor inputs to pupil control. *J Vis* 19(9):5. DOI: 10.1167/19.9.5.

Srinivasan, S., A. Cordomi, E. Ramon, and P. Garriga. 2015. Beyond spectral tuning: Human cone visual pigments adopt different transient conformations for chromophore regeneration. *Cell Mol Life Sci* 73(6):1253–1263. DOI: 10.1007/s00018-015-2043-7.

Stabio, M. E., S. Sabbah, L. E. Quattrochi et al. 2018. The M5 cell: A color-opponent intrinsically photosensitive retinal ganglion. *Cell Neuron* 97(1):150–251. DOI: 10.1016/j. neuron.2017.11.030.

Steen, R. 2011. The color of white – color consistency with LEDs. IES-Energia. http://ies -energia.com/uploads/conhecimentoLED/Color%20consistency%20with%20LEDs.pdf (accessed January 20, 2020).

Stephenson, K. M., C. M. Schroder, G. Bertschy, and P. Bourgin. 2012. Complex interaction of circadian and non-circadian effects of light on mood: Shedding new light on an old story. *Sleep Med Rev* 16(5):445–454. DOI: 10.1016/j.smrv.2011.09.002.

Stevens, J. C., and S. S. Stevens. 1963. Brightness function: Effects of adaptation. *J Opt Soc Amer* 53(3):375–385. DOI: 10.1364/JOSA.53.000375.

Stevens, S. S. 1961. To honor Fechner and repeal his law. *Science* 133(3446):80–86.

Stringham, J. M., P. V. Garcia, P. A. Smith, L. N. McLin, and B. K. Foutch. 2011. Macular pigment and visual performance in glare: Benefits for photostress recovery, disability glare, and visual discomfort. *Invest Ophthalmol Vis Sci* 52:7406–7415. DOI: 10.1167/ iovs.10-6699.

Suffern, K. 2007. *Ray tracing from the ground up*. Boca Raton, FL: A K Peters/CRC Press.

Taniyama Y., T. Yamauchi, S. Takeuchi, and Y. Kuroda. 2015. PER1 polymorphism associated with shift work disorder. *Sleep Biol Rhythms* 13(4):342–347. DOI: 10.1111/sbr.12123.

Tashiro, T., T. Kimura Minoda, S. Kohoko, T. Ishikawa, and M. Ayama. 2011. Discomfort glare evaluation to white LEDs with different spatial arrangement. In: *Proceedings of 27th CIE Session*, July 2011, SunCity, South Afrika, 583–688.

Tashiro, T., S. Kawanobe, T. Kimura-Minoda et al. 2015. Discomfort glare for white LED light sources with different spatial arrangements, *Lighting Res Technol* 47(3):316–337. DOI: 10.1177/1477153514532122.

Tebartz-van Elst, L., M. Bach, J. Blessing et al. 2015. Normal visual acuity and electrophysiological contrast gain in adults with high-functioning autism spectrum disorder. *Front Human Neurosci* 9:460. DOI: 10.3389/fnhum.2015.00460.

TechnoTeam. 2019a. LMK 6 luminance measuring video photometer: Documentation and instruction manual of the TechnoTeam's LMK photometer system. http://www.technoteam.de/ (accessed December 12, 2019).

TechnoTeam. 2019b. LMK Mobile Air. Documentation and instruction manual of the TechnoTeam's LMK photometer system. http://www.technoteam.de/ (accessed December 12, 2019).

Thapan, K., J. Arendt, and D. Skene. 2001. An action spectrum for melatonin suppression: Evidence for a novel non-rod, non-cone photoreceptor system in humans. *J Physiol* 535(1):261–267. DOI: 10.1111/j.1469-7793.2001.t01-1-00261.x.

Tomioka, N. H., H. Yasuda, H. Miyamoto et al. 2014. Elfn1 recruits presynaptic mGluR7 in trans and its loss results in seizures. *Nat Commun* 5:4501. DOI: 10.1038/ncomms5501.

Tonetti, L., and V. Natale. 2019. Effects of a single short exposure to blue light on cognitive performance. *Chronobiol Int* 36:5, 725–732, DOI: 10.1080/07420528.2019.1593191.

Tsujimura, S., K. Ukai, D. Ohama, A. Nuruki, and K. Yunokuchi. 2010. Contribution of human melanopsin retinal ganglion cells to steady–state pupil response. *Proc Biol Sci* 277(1693): 2485–2492. DOI: 10.1098/rspb.2010.0330.

Tsukamoto, Y., and N. Omi. 2013.Functional allocation of synaptic contacts in microcircuits from rods via rod bipolar to AII amacrine cells in the mouse retina. *J Comp Neurol* 521(15):3541–3555. DOI: 10.1002/cne.23370.

Tsukamoto, Y., and N. Omi. 2017. Classification of mouse retinal bipolar cells: Type-specific connectivity with special reference to rod-driven AII amacrine pathways. *Front Neuroanat* 11:92. DOI: 10.3389/fnana.2017.00092.

Udovičić, L. L. L. A.Price, and M. Khazova. 2019. Light and blue-light exposures of day workers in summer and winter. In: *Proceedings of the 29th CIE Session*, 14–22 June, 2019, Washington DC, 105–113. DOI: 10.25039/x46.2019.OP19.

Ueno, A, Y. Omori, Y. Sugita et al. 2018. Lrit1, a retinal transmembrane protein, regulates selective synapse formation in cone photoreceptor cells and visual acuity. *Cell Rep* 22(13):3548–3561. DOI: 10.1016/j.celrep.2018.03.007.

van Bommel, W. J. M., G. J. van den Beld, and M. H. F. van Ooyen. 2002. *Industrial lighting and productivity*. Philips Lighting. https://hosting.iar.unicamp.br/lab/luz/ld/Arquitet ural/interiores/ilumina%e7%e3o%20industrial/industrial_lighting_and_productivity %5b1%5d.pdf (accessed November 21, 2019).

van Bommel, W. J. M., and G. J. van den Beld. 2004. Lighting for work - a review of visual and biological effects. *Lighting Res Technol* 36(4):255–269. DOI: 10.1191/1365782804li122oa.

van Bommel, W. J. M. 2006a. Non-visual biological effect of lighting and the practical meaning for lighting for work. *Appl Ergon* 37:461–466. DOI: 10.1016/j.apergo.2006.04.009.

van Bommel, W. J. M. 2006b. Dynamic lighting at work – both in level and colour, In: *Proceedings of 2nd CIE Symposium Ottawa, Canada* 7–8 September, 2006.

van Bommel, W. J. M. 2019. *Interior lighting: Fundamentals, technology and application.* Cham, Switzerland: Springer Nature.

van den Beld, G. J. 2002. Healthy lighting, recommendations for workers. In: *Symposium healthy lighting at work and at home.* Eindhoven, the Netherlands: Eindhoven University of Technology.

van Derlofske, J. F., A. Bierman, M. S. Rea et al. 2000. Design and optimization of a retinal exposure detector, *Proc SPIE Int Soc Opt Eng* 4092:60–70.

van Lieshout-van Dal, E., L. J. A. E. Snaphaan, and I. M. B. Bongers. 2019. Biodynamic lighting effects on the sleep pattern of people with dementia. *Build Environ* 150:245–253, DOI: 10.1016/j.buildenv.2019.01.010.

van Nes, F. S., and M. A. Bouman. 1967. Spatial modulation transfer in the human eye. *J Opt Soc Am* 57(3):401–406. DOI: 10.1364/JOSA.57.000401.

van Ooyen, M. H. F., J. A. C. van de Weijgert, and S. H. A. Begemann. 1987. Preferred luminances in offices. *J Illum Eng Soc* 16(2):152–156. DOI: 10.1080/00994480.1987.1074 8695.

Vandewalle G., C. Schmidt, G. Albouy et al. 2007. Brain responses to violet, blue, and green monochromatic light exposures in humans: Prominent role of blue light and the brainstem. *PLoS One* 2(11):e1247. DOI: 10.1371/journal.pone.0001247.

Vartanian, G. V., B. Y. Li, A. P. Chervenak et al. 2015. Melatonin suppression by light in humans is more sensitive than previously reported. *J Biol Rhythms* 30(4):351–354. DOI: 10.1177/0748730415585413.

Ventura, A. L. M., A. Dos Santos-Rodrigues, C. H. Mitchell, and M. P. Faillace. 2019. Purinergic signaling in the retina: From development to disease. *Brain Res Bull* 151:92–108. DOI: 10.1016/j.brainresbull.2018.10.016.

Vicente, J., P. Laguna, A. Bartra, and R. Bailón. 2016. Drowsiness detection using heart rate variability. *Med Biol Eng Comput* 54(6):927–937. DOI: 10.1007/s11517-015-1448-7.

Viénot, F., M. L. Durand, and E. Mahler. 2009. Kruithof's rule revisited using LED illumination, *J Mod Opt* 56(13):1433–1446. DOI: 10.1080/09500340903151278.

Viney, T. J., K. Balint, D. Hillier et al. 2007. Local retinal circuits of melanopsin-containing ganglion cells identified by transsynaptic viral tracing. *Curr Biol* 17(11):981–988. DOI: 10.1016/j.cub.2007.04.058.

Wagdy, A., V. Garcia-Hansen, M. Elhenawy et al. 2019. Open-plan glare evaluator (OGE): A new glare prediction model for open-plan offices using machine learning algorithms. arXiv:1910.05594.

Wahnschaffe, A., C. Nowozin, S. Haedel et al. 2017. Implementation of dynamic lighting in a nursing home. Impact on agitation but not on rest-activity patterns. *Curr Alzheimer Res* 14(10):1076–1083. DOI: 10.2174/1567205014666170608092411.

Walmsley, L., L. Hanna, J. Mouland et al. 2015. Colour as a signal for entraining the mammalian circadian clock. *PLoS Biol* 13(4):e1002127. DOI: 10.1371/journal.pbio.1002127.

Ward G. 2019. High dynamic range image encodings. http://www.anyhere.com/gward/hdrenc/ (accessed December 12, 2019).

Webster 2019. https://www.merriam-webster.com/dictionary/light (accessed December 12, 2019).

Wei M., M. Royer, and H. P. Huang. 2019. Perceived colour fidelity under LEDs with similar Rf but different Ra. *Lighting Res Technol* 51(6):858–869. DOI: 10.1177/1477153519825997.

WELL [The WELL Building Standard]. 2019a. WELL v2, Q3 2019 version.

WELL [The WELL Building Standard]. 2019b. Feature 54: Circadian lighting design, WELL v2, Q3 2019 version. https://standard.wellcertified.com/light/circadian-lighting-design (accessed October 2, 2019).

WELL calculator [The WELL Building Standard]. 2019. WELL calculator of melanopic ratio. https://www.google.com/url?sa=t&rct=j&q=&esrc=s&source=web&cd=6&ved=2a hUKEwjP49W_or3lAhX-xMQBHTePDCgQFjAFegQIABAC&url=https%3A%2F% 2Fstandard.wellcertified.com%2Fsites%2Fdefault%2Ffiles%2FMelanopic%2520Ratio .xlsx&usg=AOvVaw2d1eSd9zZHHkSeQhG3bpQB (accessed October 2, 2019).

WELL optimization [The WELL Building Standard]. 2019. Optimization. Circadian lighting design, 2019. https://v2.wellcertified.com/v/en/light/feature/3 (accessed October 2, 2019).

Westheimer, G. 1964. Pupil size and visual resolution. *Vision Res* 4(1–2):39–45. DOI: 10.1016/0042-6989(64)90030-6.

Westland, S., Q. Pan, and S. Lee. 2017. A review of the effect of colour and light on non-image function in humans. *Color Technol* 133(5):349–361. DOI: 10.1111/cote.12289.

Weston, H. C. 1953. The relation between illumination and visual performance. Reprint: Industrial Health Research Board Report No. 87 (1945). London: HMSO, and Joint Report (1935). Medical Research Council.

Weston, H. C. 1961. Rationally recommended illuminance levels. *Trans Illum Eng Soc* 26(1):1–16.

White M. D., S. Ancoli-Israel, and R. R. Wilson. 2013. Senior living environments: Evidence-based lighting design strategies. *Herd* 7(1):60–78. DOI: 10.1177%2F193758671300700106.

Whittle, P. 1994. The psychophysics of contrast brightness. In: *Lightness, brightness, and transparency*, ed. A. L. Gilchrist. Hillsdale: Lawrence Erlbaum Associates Inc, 35–110.

WHO [World Health Organization]. 2020. Constitution. https://www.who.int/about/who-we -are/constitution (accessed January 25, 2020).

Wickwire, E. M., J. Geiger-Brown, S. M. Scharf, and C. L. Drake. 2017. Shift work and shift work sleep disorder: Clinical and organizational perspectives. *Chest* 151(5):1156–1172. DOI: 10.1016/j.chest.2016.12.007.

Wolkowitz, O. M., H. Burke, E. S. Epel, and V. I. Reus. 2009. Glucocorticoids: Mood, memory, and mechanisms. *Ann NY Acad Sci* 1179:19–40. DOI: 10.1111/j.1749-6632.2009.04980.x.

Wolska, A., and M. Śwituła. 1999. Luminance of the surround and visual fatigue of VDT operators. *Int J Occup Saf Ergon* 5(4):553–581.

Wolska, A. 2013. Glare as a specific factor in the working environment. *Przegląd Elektrotechniczny* 89(1a): 142–144.

Wolska, A., and D. Sawicki. 2013. Comparison of discomfort glare evaluation using different techniques. In: *Proceedings of the 12th European Lighting Conference Lux Europa 2013*, Cracow, Poland.

Wolska, A., and D. Sawicki. 2014. Evaluation of discomfort glare in the 50+ elderly: Experimental study. *Int J Occup Med Env* 27(3): 444–459. DOI: 10.2478/ s13382-014-0257-9.

Wolska, A., and D. Sawicki. 2016. The luminance ratio of light sources and background as a crucial factor in glare index determination – simulation analysis. In: *Proceedings of 13th Selected Issues of Electrical Engineering and Electronics (WZEE)*, 4–8, May, Rzeszow, Poland. DOI: 10.1109/WZEE.2016.7800243.

Wolska, A., D. Sawicki, K. Nowak, M. Wiselka, and M. Kołodziej. 2018. Method of acute alertness level evaluation after exposure to blue and red light (based on EEG): Technical aspects. In: *Proceedings of the 6th International Congress on Neurotechnology, Electronics and Informatics NEUROTECHNIX 2018*, 19–21 September, Seville, Spain, 53–60. DOI: 10.5220/0006922500530060.

Wolska, A., D. Sawicki, M. Kołodziej, M. Wisełka, and K. Nowak. 2019. Which EEG electrodes should be considered for alertness assessment? In: *Proceedings of the International Conference on Computer-Human Interaction Research and Applications CHIRA 2019*, 20–21 September, Vienna, Austria, 40–49. DOI: 10.5220/0008168600400049.

Wolska, A., and D. Sawicki. 2020. Practical application of HDRI for discomfort glare assessment at indoor workplaces. *Measurement* 151:107179. DOI: 10.1016/j.measurement.2019.107179.

Wong, K. Y., F. A. Dunn, D. M. Graham, and D. M. Berson. 2007. Synaptic influences on rat ganglion-cell photoreceptors. *J Physiol London* 582(1):279–296. DOI: 10.1113/jphysiol.2007.133751.

Wostyn, P. 2020. Retinal nerve fiber layer thinning in chronic fatigue syndrome as a possible ocular biomarker of underlying glymphatic system dysfunction. *Med Hypotheses* 134:109416. DOI: 10.1016/j.mehy.2019.109416.

Wright, M. 2016. DOE publishes report on the accuracy of flicker meters in characterizing LED-based lighting. *LEDs Magazine*. https://www.lcdsmagazinc.com/architcctural-lighting/retail-hospitality/article/16697059/doe-publishes-report-on-the-accuracy-of-flicker-meters-in-characterizing-ledbased-lighting (accessed October 11, 2019).

Wright, M. 2018. Researchers present circadian metrics and health impact of LED light at HCL conference. *LEDs Magazine*. https://www.ledsmagazine.com/manufNicollacturing-services-testing/substrates-wafers/article/16695782/researchers-present-circadian-metrics-and-health-impact-of-led-light-at-hcl-conference-magazine (accessed October 11, 2019).

Wyszecki, G., and W. S. Stiles. 1982. *Color science: Concepts and methods, quantitative data and formulae*. 2nd ed. New York: John Wiley & Sons.

Xu, Y., C. Orlandi, Y. Cao et al. 2016. The TRPM1 channel in ON-bipolar cells is gated by both the α and the βγ subunits of the G-protein. *Go Sci Rep* 6:20940. DOI: 10.1038/srep20940.

Xu, Q., and C. P. Lang. 2018. Revisiting the alerting effect of light: A systemic review. *Sleep Med Rev* 41:39–49.

Yuan, X., C. Zhu, M. Wang et al. 2018. Night shift work increases the risks of multiple primary cancers in women: A systematic review and meta-analysis of 61 articles. *Cancer Epidemiol Biomarkers Prev* 27(1):25–40. DOI: 10.1158/1055-9965.EPI-17-0221.

Zele, A. J., B. Feigl, P. Adhikari, M. L. Maynard, and D. Cao. 2018a. Melanopsin photoreception contributes to human visual detection, temporal and colour processing. *Sci Rep* 8:3842. DOI: 10.1038/s41598-018-22197-w.

Zele, A. J., P. Adhikari, B. Feigl, and D. Cao. 2018b. Cone and melanopsin contributions to human brightness estimation: Reply. *J Opt Soc Am A Opt Image Sci Vis* 35(10):1783. DOI: 10.1364/JOSAA.35.001783.

Zhao, J., G. Warman, and J. Cheeseman. 2019. The functional changes of the circadian system organization in aging. *Ageing Res Rev* 52:64–71. DOI: 10.1016/j.arr.2019.04.006.

Index

A

Accommodation, 7, 20, 43
Adaptation, 28, 30, 33, 43
    chromatic, 48
    circadian rhythm, 122, 125, 133, 139
    dark, 17, 20
    luminance, 57, 65
Age-related macular degeneration (AMD), 119, 149–150
Alerting stimulus, 121–122, 127
Alertness, 1, 32–33, 119, 126, 151, 169–170
    acute, 109
    assessment, 135–137, 168, 170–173
    light impact, 96–105, 125–126, 139, 144, 169–170
        blue light, 33, 100, 106, 125–127, 145
        red light, 109, 126–127
AMD, see Age-related macular degeneration
Anxiety, 26, 27

B

Blue light, 3, 92, 149
    anxiety, 26, 27
    component in the spectrum, 74, 92, 125–126, 138
    hazard, 3, 75, 149–150, 152
    impact on alertness, 29, 33, 125–126, 168
    impact on circadian system, 92, 106, 126
    impact on melatonin suppression, 3, 27, 168
    impact on cognitive performance, 145, 168
    impact on creative performance, 145–146
    impact on EEG signal, 170, 171, 172
    impact on sensory trigeminal system, 29
    stimuli, 17, 29–30, 33
Brain activity, 32, 168
    assessment, 167
    recognition, 168

C

CCR, see constant current reduction under LED
CCT, see correlated color temperature under Color
CFF, see critical fusion frequency under Flicker
CFS, see Chronic Fatigue Syndrome under Fatigue
Chromaticity diagram, 48–49, 75
Chronotype, 33, 51, 97, 120, 142, 152
    evening (eveningness), 117
    morning (morningness), 117, 135
    undifferentiated, 135

Circadian, 77–94, 105–109, 125, 144, 150–151
    amplitude, 119–120
    clock, 2, 27, 102–103, 115, 117
        gene expression, 29–30, 117
    dysregulation, 115
    misalignment, 115, 139
    mis-entrainment, 95
    phase, 1, 4
        advance, 31, 102, 119
        delay, 31, 98, 119
        shift, 28–31, 47, 95, 100, 112, 119–127
    photoentrainment, 2, 24–25, 29, 33
    preference, 117
    rhythm, 3, 29–32, 118–125, 141
    stimulation, 106, 145, 147
Circadian metrics, 77–94
    α-opic action spectra, 78, 93, 131
    α-opic metrics, 77–80
        L-cone-opic, 77
        M-cone-opic, 77
        S-cone-opic, 77
        melanopic (ipRGC), 77
        rhodopic, 77
    α-opic-radiant flux, 79
    α-opic efficacy of luminous radiation, 79–81
    α-opic irradiance, 79, 80, 131
    α-opic radiance, 79
    α-opic efficacy of luminous radiation for daylight (D65), 79
    α-opic equivalent daylight (D65) luminance, 79
    α-opic equivalent daylight (D65) illuminance, 79, 80
    α-opic equivalent daylight (D65) efficacy ratio, 79, 80
    circadian action factor, 88
    circadian efficiency, 89
    circadian light (CLA), 81–85
    circadian potency, 91
    circadian stimulus (CS), 81–85
    effective watts, 92
    equivalent melanopic lux (EML), 86–88, 130, 132, 143, 144
        calculation, 39, 87
        comparison with CS, 110–113
        determination, 86–87, 131, 165–166
        origin, 39
        values for lighting design, 109–110, 125, 130–133, 144–145
    melanopic-photopic ratio, 77, 92–93, 148